OXFORD IB DIPLOMA PROGRA

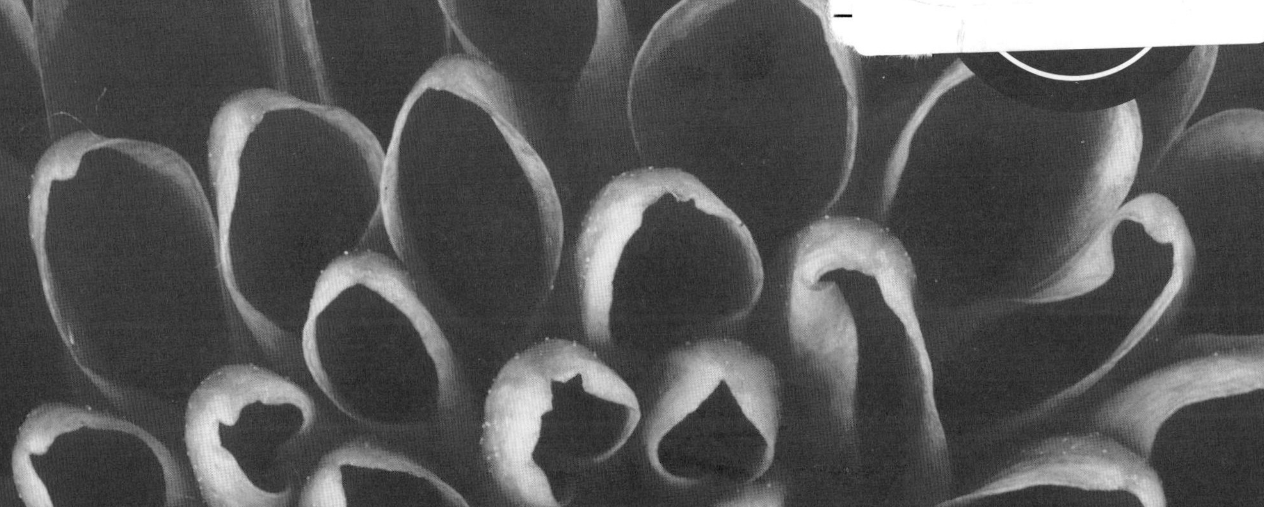

MATHEMATICS HIGHER LEVEL: DISCRETE MATHEMATICS

COURSE COMPANION

Josip Harcet
Lorraine Heinrichs
Palmira Mariz S
Marlene Torres-

OXFORD
UNIVERSITY PRESS

Great Clarendon Street, Oxford, OX2 6DP, United Kingdom

Oxford University Press is a department of the University of Oxford. It furthers the University's objective of excellence in research, scholarship, and education by publishing worldwide. Oxford is a registered trade mark of Oxford University Press in the UK and in certain other countries

First published in 2016

British Library Cataloguing in Publication Data

Data available

ISBN 978-0-19-830487-6
10 9 8 7 6 5 4 3 2 1

Paper used in the production of this book is a natural, recyclable product made from wood grown in sustainable forests.
The manufacturing process conforms to the environmental regulations of the country of origin.

Printed in Great Britain

Acknowledgements

The publisher would like to thank the following for permission to reproduce photographs:

Cover: Clive Nichols/Photo Library; **p3:** Carl Court/Stringer/Getty Images; **p4:** Dim Dimich/Shutterstock; **p20:** Royal Astronomical Society/Science Photo Library; **p41:** rook76/Shutterstock; **p42:** ©Photo Researchers/Mary Evans Picture Library; **p53:** Anton Kuba/Shutterstock; **p77:** Sergey Nivens/Shutterstock; **p85:** Antonio De Azevedo Negrão/Dreamstime; **p103:** Universal Images Group Limited/Alamy; **p124t:** gaultiero boffi/Shutterstock; **p124b:** Beto Chagas/Shutterstock; **p131:** © OpenStreetMap contributors; **p141:** Oleksiy Mark/Shutterstock; **p152:** raphael_christinat/iStock;

Course Companion definition

The IB Diploma Programme Course Companions are resource materials designed to support students throughout their two-year Diploma Programme course of study in a particular subject. They will help students gain an understanding of what is expected from the study of an IB Diploma Programme subject while presenting content in a way that illustrates the purpose and aims of the IB. They reflect the philosophy and approach of the IB and encourage a deep understanding of each subject by making connections to wider issues and providing opportunities for critical thinking.

The books mirror the IB philosophy of viewing the curriculum in terms of a whole-course approach; the use of a wide range of resources, international mindedness, the IB learner profile and the IB Diploma Programme core requirements, theory of knowledge, the extended essay, and creativity, activity, service (CAS).

Each book can be used in conjunction with other materials and indeed, students of the IB are required and encouraged to draw conclusions from a variety of resources. Suggestions for additional and further reading are given in each book and suggestions for how to extend research are provided.

In addition, the Course Companions provide advice and guidance on the specific course assessment requirements and on academic honesty protocol. They are distinctive and authoritative without being prescriptive.

IB mission statement

The International Baccalaureate aims to develop inquiring, knowledgable and caring young people who help to create a better and more peaceful world through intercultural understanding and respect.

To this end the IB works with schools, governments and international organizations to develop challenging programmes of international education and rigorous assessment.

These programmes encourage students across the world to become active, compassionate, and lifelong learners who understand that other people, with their differences, can also be right.

The IB learner Profile

The aim of all IB programmes is to develop internationally minded people who, recognizing their common humanity and shared guardianship of the planet, help to create a better and more peaceful world. IB learners strive to be:

Inquirers They develop their natural curiosity. They acquire the skills necessary to conduct inquiry and research and show independence in learning. They actively enjoy learning and this love of learning will be sustained throughout their lives.

Knowledgable They explore concepts, ideas, and issues that have local and global significance. In so doing, they acquire in-depth knowledge and develop understanding across a broad and balanced range of disciplines.

Thinkers They exercise initiative in applying thinking skills critically and creatively to recognize and approach complex problems, and make reasoned, ethical decisions.

Communicators They understand and express ideas and information confidently and creatively in more than one language and in a variety of modes of communication. They work effectively and willingly in collaboration with others.

Principled They act with integrity and honesty, with a strong sense of fairness, justice, and respect for the dignity of the individual, groups, and communities. They take responsibility for their own actions and the consequences that accompany them.

Open-minded They understand and appreciate their own cultures and personal histories, and are open to the perspectives, values, and traditions of other individuals and communities. They are accustomed to seeking and evaluating a range of points of view, and are willing to grow from the experience.

Caring They show empathy, compassion, and respect towards the needs and feelings of others. They have a personal commitment to service, and act to make a positive difference to the lives of others and to the environment.

Risk-takers They approach unfamiliar situations and uncertainty with courage and forethought, and have the independence of spirit to explore new roles, ideas, and strategies. They are brave and articulate in defending their beliefs.

Balanced They understand the importance of intellectual, physical, and emotional balance to achieve personal well-being for themselves and others.

Reflective They give thoughtful consideration to their own learning and experience. They are able to assess and understand their strengths and limitations in order to support their learning and personal development.

A note on academic honesty

It is of vital importance to acknowledge and appropriately credit the owners of information when that information is used in your work. After all, owners of ideas (intellectual property) have property rights. To have an authentic piece of work, it must be based on your individual and original ideas with the work of others fully acknowledged. Therefore, all assignments, written or oral, completed for assessment must use your own language and expression. Where sources are used or referred to, whether in the form of direct quotation or paraphrase, such sources must be appropriately acknowledged.

How do I acknowledge the work of others?

The way that you acknowledge that you have used the ideas of other people is through the use of footnotes and bibliographies.

Footnotes (placed at the bottom of a page) or endnotes (placed at the end of a document) are to be provided when you quote or paraphrase from another document, or closely summarize the information provided in another document. You do not need to provide a footnote for information that is part of a 'body of knowledge'. That is, definitions do not need to be footnoted as they are part of the assumed knowledge.

Bibliographies should include a formal list of the resources that you used in your work. The listing should include all resources, including books, magazines, newspaper articles, Internet-based resources, CDs and works of art. 'Formal' means that you should use one of the several accepted forms of presentation. You must provide full information as to how a reader or viewer of your work can find the same information. A bibliography is compulsory in the extended essay.

What constitutes misconduct?

Misconduct is behaviour that results in, or may result in, you or any student gaining an unfair advantage in one or more assessment component. Misconduct includes plagiarism and collusion.

Plagiarism is defined as the representation of the ideas or work of another person as your own. The following are some of the ways to avoid plagiarism:

- Words and ideas of another person used to support one's arguments must be acknowledged.
- Passages that are quoted verbatim must be enclosed within quotation marks and acknowledged.
- CD-ROMs, email messages, web sites on the Internet, and any other electronic media must be treated in the same way as books and journals.
- The sources of all photographs, maps, illustrations, computer programs, data, graphs, audio-visual, and similar material must be acknowledged if they are not your own work.
- Words of art, whether music, film, dance, theatre arts, or visual arts, and where the creative use of a part of a work takes place, must be acknowledged.

Collusion is defined as supporting misconduct by another student. This includes:

- allowing your work to be copied or submitted for assessment by another student
- duplicating work for different assessment components and/or diploma requirements.

Other forms of misconduct include any action that gives you an unfair advantage or affects the results of another student. Examples include, taking unauthorized material into an examination room, misconduct during an examination, and falsifying a CAS record.

About the book

The new syllabus for Mathematics Higher Level Option: Discrete Mathematics is thoroughly covered in this book. Each chapter is divided into lesson-size sections with the following features:

The Course Companion will guide you through the latest curriculum with full coverage of all topics and the new internal assessment. The emphasis is placed on the development and improved understanding of mathematical concepts and their real life application as well as proficiency in problem solving and critical thinking. The Course Companion denotes questions that would be suitable for examination practice and those where a GDC may be used.

Questions are designed to increase in difficulty, strengthen analytical skills and build confidence through understanding.

Where appropriate the solutions to examples are given in the style of a graphics display calculator.

Mathematics education is a growing, ever changing entity. The contextual, technology integrated approach enables students to become adaptable, lifelong learners.

Note: US spelling has been used, with IB style for mathematical terms.

About the authors

Lorraine Heinrichs has been teaching mathematics for 30 years and IB mathematics for the past 16 years at Bonn International School. She has been the IB DP coordinator since 2002. During this time she has also been senior moderator for HL Internal Assessment and workshop leader of the IB; she was also a member of the curriculum review team.

Palmira Mariz Seiler has been teaching mathematics for over 25 years. She joined the IB community in 2001 as a teacher at the Vienna International School and since then has also worked as Internal Assessment moderator in curriculum review working groups and as a workshop leader and deputy chief examiner for HL mathematics. Currently she teaches at Colegio Anglo Colombiano in Bogota, Colombia.

Marlene Torres-Skoumal has taught IB mathematics for over 30 years. During this time, she has enjoyed various roles with the IB, including deputy chief examiner for HL, senior moderator for Internal Assessment, calculator forum moderator, workshop leader, and a member of several curriculum review teams.

Josip Harcet has been involved with and teaching the IB programme since 1992. He has served as a curriculum review member, deputy chief examiner for Further Mathematics, assistant IA examiner and senior examiner for Mathematics HL as well as a workshop leader since 1998.

Contents

Chapter 1 Making sense of numbers **2**
Introduction A brief journey through different number systems 3
1.1 Number systems and bases 4
1.2 Integers, prime numbers, factors and divisors 13
Diophantus of Alexandria 20
Linear Diophantine equations 21
Prime numbers 26
1.3 Strong mathematical induction 30
1.4 The Fundamental Theorem of Arithmetic and least common multiples 33
Review exercise 37

Chapter 2 Modular arithmetic and its applications **40**
Introduction From Gauss to cryptography 41
2.1 Congruence modulo n 42
2.2 Modular inverses and linear congruences 48
2.3 The Pigeonhole Principle 53
2.4 The Chinese Remainder Theorem or systems of linear congruences 57
2.5 Using cycles for powers modulo n and Fermat's Little Theorem 64
Review exercise 72

Chapter 3 Recursive patterns **76**
Introduction Modelling and solving problems using sequences 77
3.1 Recurrence relations 78
3.2 Solution of first-degree linear recurrence relations and applications to counting problems 83
3.3 Modelling with first-degree recurrence relations 89
Financial problems 89
Loans and amortizations 90
Investments and compound interest 91
Games and probability problems 92
3.4 Second-degree linear homogeneous recurrence relations with constant coefficients 94
Review exercise 99

Chapter 4 From folk puzzles to a new branch of mathematics **102**
Introduction Introduction to graph theory 103
4.1 Terminology and classification of graphs 104
What is a graph and what are its elements? 104
4.2 Classification of graphs 108
Weighted graphs 108
Directed graphs 109
Simple graphs 109
Connected graphs 110
Trees 111
Complete graphs 112
Bipartite graphs 113
4.3 Different representations of the same graph 115
4.4 Planar graphs 118
Spanning trees 119
Complements of graphs 119
Euler relation for planar graphs 120
Real life application – The soccer ball 124
4.5 Hamiltonian cycles 126
4.6 Eulerian circuits and trails 129
Review exercise 133

Chapter 5 Applications of Graph Theory **140**
Introduction Further algorithms and methods 141
5.1 Graph algorithms: Kruskal's and Dijkstra's 142
Minimum Connector Problems 142
Shortest Path Problems 146
Dijkstra's Algorithm 147
Limitation of Dijkstra's Algorithm 148
5.2 Chinese postman problem 149
Chinese Postman algorithm 151
5.3 Travelling Salesman Problem 153
The Nearest Neighbour Algorithm for upper bound 154
Deleted vertex algorithm for lower bound 155
Review exercise 160

Answers **165**

Index **177**

1 Making sense of numbers

CHAPTER OBJECTIVES:

10.1 Strong induction.

10.2 Division and Euclidean algorithms. The greatest common divisor, gcd(*a*, *b*), and the least common multiple, lcm(*a*, *b*), of integers *a* and *b*.

10.3 Linear Diophantine equations $ax + by = c$.

10.5 Representation of integers in different bases.

Before you start

You should know how to:

1 Prove statements directly by factorization, e.g. $n^2 + 9n + 20$ is an even number for all $n \in \mathbb{Z}^+$ since $n^2 + 9n + 20 \equiv (n + 4)(n + 5)$ which is a product of two consecutive numbers. For any two consecutive integers, one of them must be even, and $n^2 + 9n + 20$ is the product of an odd and an even positive integer making it even.

2 Prove statements using mathematical induction, e.g. Prove the statement

$P_n : \sum_{i=1}^{n} i(2^i) = 2 + (n - 1) \times 2^{n+1}$

Proof: When $n = 1$,
LHS $= 1 \times 2^1 = 2$, RHS $= 2 + 0 = 2$.
Therefore P_1 is true.
Assume that P_k is true for some $k \geq 1$.
i.e. $\sum_{i=1}^{k} i(2^i) = 2 + (k - 1) \times 2^{k+1}$
When $n = k + 1$,
$\sum_{i=1}^{k+1} i(2^i) = 2 + (k - 1) \times 2^{k+1} + (k + 1)(2^{k+1})$
$= 2 + 2^{k+1}(k - 1 + k + 1) = 2 + 2k(2^{k+1})$
$= 2 + k \times 2^{k+2}$
$\Rightarrow P_{k+1}$ is true.
Since P_1 is true and it was shown that if P_k is true then P_{k+1} is also true, it follows by the principle of mathematical induction that P_n is true for all $n \geq 1$, $n \in \mathbb{Z}^+$. Q.E.D.

Skills check:

1 **a** Show that $n^3 - n$ is divisible by 6 for all $n \in \mathbb{Z}^+$.

b Show that $n^5 - n$ is divisible by 30 for all $n \in \mathbb{Z}^+$.

2 **a** Prove the two statements in question **1** using the principle of mathematical induction.

b Using the principle of mathematical induction prove that

$$\sum_{i=1}^{n} i(i + 2) = \frac{n(n + 1)(2n + 7)}{6}$$

A brief journey through different number systems

One would think that human beings have a very good sense of numbers because of our ability to count. However, studies show that tribal people who have not developed the skill of counting have difficulty discerning beyond the quantity 4. When it comes to number sense we are not much different from other species in the animal kingdom. However, although our number sense is limited, we are all able to learn how to count and this is what makes us different.

Throughout history people have devised systems to aid keeping track of quantity. The Mesopotamians had a number system using base 60 as far back as 3400 BC and the Egyptian number system dates back to 3100 BC. The Egyptians used base 10 in their system and they had a special symbol for the different powers of 10, allowing them to count up to one million.

Here are the Egyptian symbols for the powers of ten from 10 to one million.

𓎆	𓍢	𓆼	𓂭	𓆐	𓁨
10	100	1000	10 000	100 000	1 million

In Europe, Roman numbers were used before our current number system. One of the most ancient systems is the Mayan system which was developed around 400 AD, appoximately 1000 years ahead of

European counterparts. The Mayans used base 20 in their system. The picture below illustrates some of the numbers in the Mayan system.

1 4 6 9 11 17 zero 20 23

But it was in India that the number zero was first introduced as a conceptual number and indirectly revolutionized western arithmetic many centuries later. Until then, western arithmetic used Roman numerals which made arithmetic very cumbersome.

❓ The Indian mathematician Brahmagupta came up with some rules about operations with positive numbers, negative numbers and zero.

Rules of Brahmagupta

A debt minus zero is a debt.
A fortune minus zero is a fortune.
Zero minus zero is a zero.
A debt subtracted from zero is a fortune.
The product of zero multiplied by a debt or fortune is zero.
The product of zero multiplied by zero is zero.
The product or quotient of two fortunes is one fortune.
The product or quotient of two debts is one fortune.
The product or quotient of a debt and a fortune is a debt.
The product or quotient of a fortune and a debt is a debt.

Investigation

Work out the following products:

34×11 71×11 23×11 29×11

What do you notice from your results? Work out some more products of two-digit numbers by 11 and make a conjecture.
Does your conjecture work with the product of three-digit numbers and 11?
What about four-digit or five-digit numbers, etc...?
Explain why your conjecture always works.

Repeat the steps above for multiplication by 111 and explain your results.

1.1 Number systems and bases

In discrete mathematics we are interested only in values that vary discretely, as opposed to values that vary continuously. Hence, the focus will be on variables that belong to the integers $\mathbb{Z}$, and subsets thereof, particularly the positive integers $\mathbb{Z}^+$. In this section you will look at different number bases and learn how to convert from one base to another, as well as doing some elementary arithmetic.

Since a very early age you have been using the base 10 system with the digits 0 through 9. You were taught how to understand the value of a number based on a decimal system where the value of a digit depends on where it is placed in the representation of the number. This is illustrated in the table below.

10^4	10^3	10^2	10^1	10^0
2	3	0	4	7

We can think of this number in base 10 as a polynomial in 10 with coefficients that can take values between 0 and 9 as follows:

$23\,047 = 2 \times 10^4 + 3 \times 10^3 + 0 \times 10^2 + 4 \times 10^1 + 7 \times 10^0$

i.e. $f(x) = 2x^4 + 3x^3 + 4x + 7 \Rightarrow f(10) = 23\,047$

We are so accustomed to using this system that we often do our calculations mechanically. Most probably, the reason base 10 mathematics was adopted by many civilizations is because we have 10 fingers.

? *Digitus* in Latin means a finger or a toe. In the Mayan system, all fingers and toes were used, resulting in a base 20 number system, whereas Egyptians used only fingers, hence the base 10 system. Native Greenlanders also used fingers and toes. In fact the Greenlandic word for eight translates as 'second hand three' – meaning you count the five fingers of the first hand and three fingers from the second hand. The number 14 in Greenlandic would translate directly as 'first foot four' – meaning all 10 fingers and 4 toes. So 'second foot 4' would be the number 19.

Computer systems and digital electronics use binary (base 2), octal (base 8) and hexadecimal (base 16) number systems for internal storage and the processing of data. The binary system is especially useful for representing the input and output of computer components and for memory storage locations that can be in only one of two states, on or off. This system uses only two digits: 1 and 0. The table below illustrates an example of how to understand the value of a binary number in denary.

Base ten is called either decimal or denary.

2^4	2^3	2^2	2^1	2^0
1	0	0	1	1

Once more we can think of the number as a polynomial in 2 with coefficients that can take values 0 or 1 as follows:

$10011_2 = 1 \times 2^4 + 0 \times 2^3 + 0 \times 2^2 + 1 \times 2^1 + 1 \times 2^0$

$= 16 + 2 + 1$

$= 19_{10}$

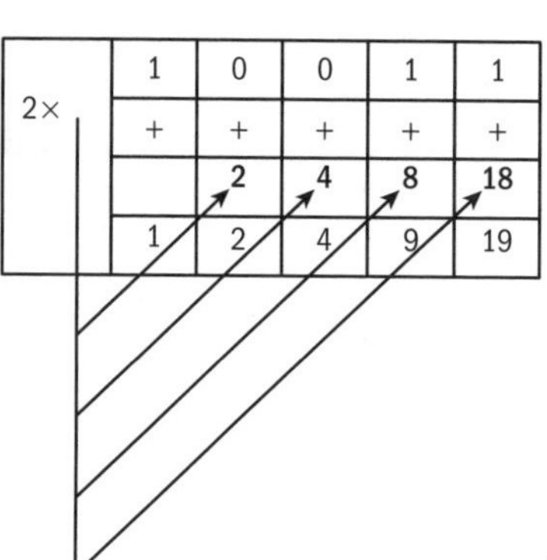

In the core book we introduced Horner's algorithm, also known as synthetic division. The table also helps us to calculate the number in base 10.

Definition

A positive integer N in base b notation is represented by $N = (d_n d_{n-1} d_{n-2} \dots d_1 d_0)_b$ where $d_i, b \in \mathbb{Z}^+, 0 \le d_i < b$.

The **value** of N in base b is given by:

$N = d_n \times b^n + d_{n-1} \times b^{n-1} + d_{n-2} \times b^{n-2} + \dots + d_1 \times b^1 + d_0 \times b^0$

The hexadecimal system is base 16. Using the definition for base b notation, a number N in hexadecimal would be represented by:

$N = d_n \times 16^n + d_{n-1} \times 16^{n-1} + d_{n-2} \times 16^{n-2} + \dots + d_1 \times 16^1 + d_0 \times 16^0$

with $d_i \in \mathbb{Z}^+, 0 \le d_i < 16$

Since we do not have 16 different digits to represent the different d_i we use letters as additional digits, so that in the hexadecimal system the digits used are 0, 1, 2, 3, 4, 5, 6, 7, 8, 9, A, B, C, D, E, F with $A_{16} = 10$, $B_{16} = 11$, $C_{16} = 12$, $D_{16} = 13$, $E_{16} = 14$, and $F_{16} = 15$.

Example 1 illustrates how to use this definition to find the value of a number from a given base to base 10.

Example 1

Convert the following numbers to base 10.

a 35072_8 **c** 1101011_2

b 4211_5 **d** $EA21B_{16}$

a $35072_8 = 3 \times 8^4 + 5 \times 8^3 + 0 \times 8^2 + 7 \times 8^1 + 2$ $= 12288 + 2560 + 56 + 2 = 14906_{10}$	*The number is in base 8, so use the definition with b = 8.*
b $4211_5 = 4 \times 5^3 + 2 \times 5^2 + 1 \times 5^1 + 1 = 556_{10}$	*Use the definition with b = 5.*
c 1101011_2 $= 1 \times 2^6 + 1 \times 2^5 + 0 \times 2^4 + 1 \times 2^3 + 0 \times 2^2 + 1 \times 2^1 + 1$ $= 107_{10}$	*Use the definition with b = 2.*
d $EA21B_{16} = 14 \times 16^4 + 10 \times 16^3 + 2 \times 16^2 + 1 \times 16^1 + 11$ $= 959003_{10}$	*Use the definition with b = 16, E = 14, A = 10 and B = 11.*

In the exam you will be required to work with bases only up to and including base 16. Remember that for bases greater than 10 the letters A, B, C.... represent the numbers 10, 11, 12... You should try to solve the questions in example 1 using synthetic division (Horner's algorithm).

In the next example you will determine the base used in an equation by using a representation of the number in terms of powers. This is sometimes called a polynomial representation of a number.

Usually we denote the base by b, but in the next example the base is called n so that you may follow the working using a GDC, as shown in the screenshot.

Example 2

When numbers are written in base n, $44^2 = 4301$. By writing down an appropriate polynomial equation, determine the value of n.

$44^2 = 4301$
$\Rightarrow (4n + 4)^2 = 4n^3 + 3n^2 + 0n + 1$
$\Rightarrow 16n^2 + 32n + 16 = 4n^3 + 3n^2 + 1$
$\Rightarrow 4n^3 - 13n^2 - 32n - 15 = 0$
$\Rightarrow n = 5$

Write down each side of the equation as a polynomial in n. Expand and simplify then solve for $n \in \mathbb{Z}^+$.

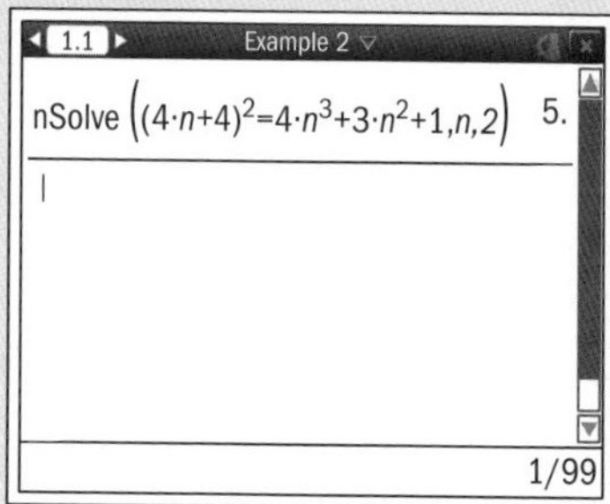

You can also use a GDC. Notice that for the first iteration we used 2, since that is the smallest base.
Note that although -1 and $-\frac{3}{4}$ are also solutions these numbers cannot be used as a base.

To change a decimal number into any base b you need to work backwards. Suppose that you want to change the number 163_{10} to base 3. Start by listing the powers of 3 and choose the largest power that is smaller than 163.

$3^2 = 9$, $3^3 = 27$, $3^4 = 81$, $3^5 = 243$

Dividing 163 by 81 you get 2 with a remainder of 1.

So $163 = 2 \times 3^4 + 0 \times 3^3 + 0 \times 3^2 + 0 \times 3^1 + 1 = 20001_3$.

Another algorithm involves successive division by the base and noting the remainders at each stage as follows:

> An *algorithm* is a step-by-step set of operations that are to be followed in calculations or problem solving.

$163 = 3 \times 54 + 1$
$54 = 3 \times 18 + 0$
$18 = 3 \times 6 + 0$
$6 = 3 \times 2 + 0$
$2 = 3 \times 0 + 2$

The remainders written from last to first give the answer 20001_3.

The algorithms used for adding and multiplying in base 10 can also be used for adding and multiplying in base b as long as you remember to convert to base b at each stage.

To add numbers, you start by adding the digits on the far right, the units digits. If the sum is less than the base you write it down. If it is greater than the base then you must carry a number. To find out what number to carry, you divide the sum by the base, write down the remainder and carry up the quotient. Repeat this in each place digit, remembering to add any carried numbers at each stage. Note that if you are adding two numbers you cannot carry more than a 1. In that case, carry a 1 if the sum is greater than or equal to the base.

The following example will illustrate the algorithm.

$372_8 + 437_8 = 1031_8$

Method:

Start with the units digit.

$2 + 7 = 9$

$9 = 1 \times 8 + 1 = 11_8$

The sum is greater than 8 so write 1 in the units digit and carry 1:

$$\begin{array}{r} \overset{1}{3}\overset{}{7}2_8 \\ +\ 437_8 \\ \hline 1 \end{array}$$

Now we have $7 + 3 + 1 = 11 = 1 \times 8 + 3 = 13_8$

Put 3 in for the 8^1 digit and carry 1:

$$\begin{array}{r} \overset{1}{3}72_8 \\ +\ 437_8 \\ \hline 31 \end{array}$$

$3 + 4 + 1 = 8 = 1 \times 8 + 0 = 10_8$

Put 0 in for the 8^2 digit and carry 1 over to the 8^3 digit:

$$\begin{array}{r} 372_8 \\ +\ 437_8 \\ \hline 1031_8 \end{array}$$

To multiply two numbers, start by multiplying the units digit first. If the product is less than the base you write it down. If it is greater than the base then you must carry. To find out what number to carry you divide the product by the base, write down the remainder and carry up the quotient. Repeat this in each place digit but remember to add any carried numbers at each stage.

The following example will illustrate the algorithm.

$253_7 \times 25_7 = 10\,351_7$

Method:

$3 \times 5 = 15 = 2 \times 7 + 1$ so write 1 in the units digit and carry 2:

$$\begin{array}{r} 2\overset{2}{5}3 \\ \times\ \ 25 \\ \hline 1 \end{array}$$

$5 \times 5 + 2 = 27 = 3 \times 7 + 6$ so write 6 for the 7^1 digit and carry 3:

$$\begin{array}{r} \overset{3}{2}53 \\ \times\ \ 25 \\ \hline 61 \end{array}$$

$5 \times 2 + 3 = 13 = 1 \times 7 + 6$ therefore write 16 for the 7^2 and 7^3 digits:

$$\begin{array}{r} 253 \\ \times\ \ 25 \\ \hline 1661 \end{array}$$

Continuing with the algorithm of multiplying by 20, place a 0 in the units digit and multiply by 2.

$2 \times 3 = 6$ and since $6 < 7$ we write it down:

$$\begin{array}{r} 253 \\ \times\ \ 25 \\ \hline 1661 \\ 60 \end{array}$$

$2 \times 5 = 10 = 1 \times 7 + 3$ so we write 3 and carry 1:

$$\begin{array}{r} \overset{1}{2}53 \\ \times\ \ 25 \\ \hline 1661 \\ 360 \end{array}$$

$2 \times 2 + 1 = 5$ and we write 5:

$$\begin{array}{r} 253 \\ \times\ \ 25 \\ \hline 1661 \\ 5360 \end{array}$$

The last step is to add in base 7:

$$\begin{array}{r} 253 \\ \times\ \ 25 \\ \hline 1661 \\ +\ 5360 \\ \hline 10351 \end{array}$$

Therefore $253_7 \times 25_7 = 10\,351_7$.

Base 7 times table	0	1	2	3	4	5	6	7
0	0							
1	0	1						
2	0	2	4					
3	0	3	6	12				
4	0	4	11	15	22			
5	0	5	13	21	26	34		
6	0	6	15	24	33	42	51	
7	0	10	20	30	40	50	60	100

The next example illustrates how multiplication in a different base can be used within exam-style questions.

Example 3

Rewrite the equality $(44_5)^2 = 4301_5$ with numbers in base 3.

Method I	
$44_5 = 4 \times 5 + 4 = 24_{10} \Rightarrow (24_{10})^2 = 576_{10}$	*Change both sides of the equation to base 10.*
$24 = 3 \times 8 + 0$ $8 = 3 \times 2 + 2$ $2 = 3 \times 0 + 2$ $\Rightarrow 24_{10} = 220_3$	*Convert 24_{10} to base 3.*
$576 = 3 \times 192 + 0$ $192 = 3 \times 64 + 0$ $64 = 3 \times 21 + 1$ $21 = 3 \times 7 + 0$ $7 = 3 \times 2 + 1$ $2 = 3 \times 0 + 2$ $\Rightarrow 576_{10} = 210\,100_3$	*Convert 576_{10} to base 3.*
Therefore the equation in base 3 becomes $(220_3)^2 = 210\,100_3$	
Method II	
$44_5 = 24_{10}$	
$24 = 3 \times 8 + 0$ $8 = 3 \times 2 + 2$ $2 = 3 \times 0 + 2$ $\Rightarrow 24_{10} = 220_3$ Therefore $44_5 = 220_3$	*Convert 24_{10} to base 3.*
220_3 $\times 220_3$ 12100_3 1210_3 210100	*Square 220_3* 22_3 $\times 2_3$ *because* $2_3 \times 2_3 = 11_3$ 121_3
Therefore in base 3, $(220_3)^2 = 210\,100_3$	

Computers store and handle binary digits, where each digit is called a bit. In order to represent hexadecimal digits we need four binary digits. Two hexadecimal digits together make up a byte. The hexadecimal system is used by programmers because each byte needs only two hexadecimal digits, and also hexadecimal numbers are more easily read by humans than binary numbers. In HTML and CSS codes, hexadecimal triplets are used to specify colours. Each of these triplets consists of two hexadecimal numbers. The table at the top of page 11 shows the numbers from 0 to 15 in denary (decimal), hexadecimal and binary.

The first 16 digits in the Base 10 (denary), Base 16 (hexidecimal) and Base 2 (binary) number systems

denary	hexadecimal	binary
0	0	0000
1	1	0001
2	2	0010
3	3	0011
4	4	0100
5	5	0101
6	6	0110
7	7	0111
8	8	1000
9	9	1001
10	A	1010
11	B	1011
12	C	1100
13	D	1101
14	E	1110
15	F	1111

? Octal (Base 8) is another numbering system, and some sources claim it was used by several native American tribes. Rather than counting fingers, as with base 10, the spaces in between the fingers were counted.

The next example shows how to prove divisibility properties.

Example 4

Let N be a positive integer written in base 6. Show that N is divisible by 5 if and only if the sum of the digits of N is divisible by 5.

$N_6 = a_n \times 6^n + a_{n-1} \times 6^{n-1} + a_{n-2} \times 6^{n-2} + \ldots + a_2 \times 6^2 + a_1 \times 6 + a_0$ where $0 \le a_i < 6$ for all $0 \le i \le n$	*Write N_6.*
$N_6 = a_n(1+5)^n + a_{n-1}(1+5)^{n-1} + a_{n-2}(1+5)^{n-2} + \ldots + a_2(1+5)^2 + a_1(1+5) + a_0$	*Substitute $6 = (1 + 5)$.*

$(1+5)^k = \binom{k}{0}5^k + \binom{k}{1}5^{k-1} + \binom{k}{2}5^{k-2} + \ldots + \binom{k}{k-1}5 + \binom{k}{k}$	*Using Binomial expansion.*
$= 5\underbrace{\left[\binom{k}{0}5^{k-1} + \binom{k}{1}5^{k-2} + \binom{k}{2}5^{k-3} + \ldots + \binom{k}{k-1}\right]}_{A_k} + 1$	*Rearrange.*
$= 5A_k + 1$ $N_6 = a_n(5A_n+1) + a_{n-1}(5A_{n-1}+1) + a_{n-2}(5A_{n-2}+1)$ $+ \ldots + a_2(5A_2+1) + a_1(5A_1+1) + a_0$ $= 5\underbrace{(a_nA_n + a_{n-1}A_{n-1} + a_{n-2}A_{n-2} + \ldots + a_2A_2 + a_1A_1)}_{B}$ $+ \underbrace{(a_n + a_{n-1} + a_{n-2} + \ldots + a_1 + a_0)}_{S}$ $= 5B + S$	*Substitute $5A_k + 1$ for $(1+5)^k$ in the expression for N_6.*
$5 \mid S \Rightarrow S = 5p, p \in \mathbb{Z}^+$ $\Rightarrow N_6 = 5(B+p)$ $\Rightarrow 5 \mid N_6$	*Show that $5 \mid S \Rightarrow 5 \mid N_6$.*
$5 \mid N_6 \Rightarrow N_6 = 5C$ $\Rightarrow 5B + S = 5C$ $\Rightarrow S = 5(C-B)$ $\Rightarrow 5 \mid S$	*Show that $5 \mid N_6 \Rightarrow 5 \mid S$.*

The result above can be generalized for any number written in base b. This is left as an exercise for you in question 10 of Exercise 1A.

Exercise 1A

1 Convert these numbers to base 10.

a 4578_9 **b** $EB7F4_{16}$ **c** $312\,201_4$

2 Convert these numbers from base 10 to the given base.

a 82 966 to base 16 **b** 73 285 to base 5 **c** 347 to base 2

3 Convert these numbers to base 6.

a 2122_3 **b** $C19_{16}$ **c** $11\,011\,101_2$

4 Work out the sums in the given base.

a $565_7 + 2154_7$ **b** $A5_{16} + B38_{16}$ **c** $110101_2 + 1011100_2$

5 Work out the products in the given base.

a $2314_5 \times 43_5$ **b** $111_2 \times 110_2$ **c** $(53_9)^2$

6 When numbers are written in base b, $25 \times 16 = 502$. By writing down an appropriate polynomial equation, determine the value of b.

7 a When numbers are written in base n, $34^2 = 2421$. By writing down an appropriate polynomial equation, determine the value of n.

b Rewrite the above equality with numbers in base 8.

8 Let N be a positive integer written in base 9. Show that N is divisible by 8 if and only if the sum of the digits of N in base 9 is divisible by 8.

9 Let N be a positive integer expressed in base 10, i.e. $N = a_n \times 10^n + a_{n-1} \times 10^{n-1} + \ldots + a_2 \times 10^2 + a_1 \times 10 + a_0$

Show that N is divisible by 4 if and only 4 divides $2a_1 + a_0$.

10 Show that if N is a positive integer written in base k, then $(k-1)$ divides N if and only if the sum of the digits of N in base k is a multiple of $(k-1)$.

1.2 Integers, prime numbers, factors and divisors

We will now look at some basic concepts of Number Theory with special emphasis on divisibility, divisors, remainders and greatest common divisors.

Leopold Kronecker (1823–1891) was a German mathematician whose work focused on Number Theory and Algebra. He was quite critical of the work done by Cantor and Weierstrass on Set Theory. He is famous for saying: *"God made the integers, all else is the work of man."*

Definition

If $a, b \in \mathbb{Z}, a \neq 0$, we say that a divides b if there exists $c \in \mathbb{Z}$ such that $b = ac$. We then say that a is a **factor** of b, and b is a **multiple** of a. The notation $a \mid b$ denotes that a divides b.

i.e. $a \mid b \Rightarrow b = ac$ where $a, b, c \in \mathbb{Z}, a \neq 0$.

Theorem 1

Let $a, b, c \in \mathbb{Z}, a \neq 0$. Then

i $a \mid b$ and $a \mid c \Rightarrow a \mid (b + c)$

ii $a \mid b \Rightarrow a \mid bc$

iii $a \mid b$ and $b \mid c \Rightarrow a \mid c$

Proof of Theorem 1:

i:

$a \mid b \Rightarrow b = am$, where $m \in \mathbb{Z}$.

$a \mid c \Rightarrow c = an$, where $n \in \mathbb{Z}$.

Therefore $b + c = am + an = a(m + n)$, where $m + n \in \mathbb{Z} \Rightarrow a \mid (b + c)$. Q.E.D.

ii:

$a \mid b \Rightarrow b = am$, where $m \in \mathbb{Z}$.

$\Rightarrow bc = amc$, where $mc \in \mathbb{Z} \Rightarrow a \mid bc$. Q.E.D.

iii:

$a \mid b \Rightarrow b = am$, where $m \in \mathbb{Z}$.

$b \mid c \Rightarrow c = nb$, where $n \in \mathbb{Z}$.

$\Rightarrow c = nb = amn$ (substituting for b)

and since $mn \in \mathbb{Z}$, we have $a \mid c$. Q.E.D.

Part iii of theorem 1 is called the transitive property.

It is left as an exercise for you to prove the following properties:

- $a \mid a$ for all $a \in \mathbb{Z}$, $a \neq 0$.
- If $a \mid b$ and $b \mid a$, $a, b \in Z$, $b \neq 0$, then a and b are equal or opposite.

Two numbers having the same magnitude but different signs are said to be opposite numbers.

An integer may or may not be divisible by another integer. When an integer is divided by another integer you always obtain a quotient and a remainder (though the remainder may be 0). You have been using this in your arithmetic calculations, possibly without realizing how important the division algorithm is. Some examples:

When 113 is divided by 11 the quotient is 10 and the remainder is 3 because $113 = 11 \times 10 + 3$.

When -42 is divided by 11 the quotient is -4 and the remainder is 2 because $-42 = 11 \times (-4) + 2$. (*Note that the remainder cannot be negative and if the number is negative then the quotient is negative.*)

When 104 is divided by -5 the quotient is -20 and the remainder is 4 because $104 = (-5) \times (-20) + 4$.

Definition

A relation $\leq$ on a set S is *totally ordered* if and only if for all a, b, c, in S the following hold

- either $a \leq b$ or $b \leq a$
- $a \leq b$ and $b \leq a \Rightarrow a = b$
- $a \leq b$ and $b \leq c \Rightarrow a \leq c$

A **well-ordered** relation on a set S is one that is totally ordered, and every non-empty subset of S contains a least element.

The Well-Ordering Principle

Every non-empty subset of the positive integers has a least element. We say that the positive integers are **well-ordered**.

It is clear that $\mathbb{Z}^+$ is well-ordered because every subset of $\mathbb{Z}^+$ contains a least element. For example, $E = \{k \mid k = 2n, n \in \mathbb{Z}^+\}$ is a subset of $\mathbb{Z}^+$. It is the set of all even positive integers with least element 2. The least element of $\mathbb{Z}^+$ is 1.

The set $\mathbb{N}$ is also well-ordered since it it has a least element of 0 and any subset of $\mathbb{N}$ also has a least element.
For example, $S = \{k \mid k = 5n, n \in \mathbb{Z}^+\} \Rightarrow S \subset \mathbb{N}$ with least element 5.

If we consider $[0, \infty[$ then this is not well-ordered. Although it has a least element, 0, the set $]0, \infty[\subset [0, \infty[$ but it has no least element.

Theorem 2: The division theorem

Let $a \in \mathbb{Z}$ and $d \in \mathbb{Z}^+$. Then there are unique integers q and r, $0 \leq r < d$, such that $a = dq + r$. We call d the divisor, q the quotient and r the remainder.

Proof:

You will not be required to prove Theorem 2 in the examination.

The proof is divided into two parts. In the first part we need to show that there are two integers q and r, $0 \leq r < d$, such that $a = dq + r$.
In the second part we must show that these integers are unique.

i Existence of q and r:

If $d \mid a$ then $a = nd$ and $r = 0$.

If $d \nmid a$, let $S = \{a - kd \mid k \in \mathbb{Z}, a - kd > 0\}$.

$d \nmid a$ means d does not divide a.

If $a > 0$ and $k = 0$ we have $a \in S$ so we know that $S \neq \varnothing$.

For $a < 0$ let $k = a - 1$ so that $a - kd = a - (a - 1)d = a(1 - d) + d$ where $1 - d \leq 0$ since $d \geq 1$. Then $a - kd > 0$ and once more $S \neq \varnothing$. Hence, for all $a \in \mathbb{Z}$, S is a non-empty subset of $\mathbb{Z}^+$ and by the Well-Ordering Principle it must have a least element r such that $0 < r = a - qd$ for some $q \in \mathbb{Z}$. If $r = d$ then $d = a - qd \Rightarrow a = d(q + 1) \Rightarrow d \mid a$ which is a contradiction since the initial condition was that $d \nmid a$.

If $r > d$ we can say that $r = d + c$ where $c \in \mathbb{Z}^+$ and $d + c = a - qd \Rightarrow a = d(1 + q) + c$ which leads to $c \in S$ and $c < r$. This is another contradiction since we said that r is the least element in S. Therefore we know that q and r exist such that $r < d$ and $a = qd + r$. Q.E.D.

ii Uniqueness of q and r:

Suppose there exist q_1, r_1, q_2, r_2 such that $q_1 d + r_1 = a = q_2 d + r_2$
Then it follows that $|q_1 - q_2|d = |r_2 - r_1| < d$ because we know that $0 \leq r_1, r_2 < d$. If $q_1 \neq q_2$ then $|q_1 - q_2|d > d$ which is a contradiction.
Therefore $q_1 = q_2$ and hence $r_1 = r_2$. Q.E.D.

Definition

Given that $a, b \in \mathbb{Z}^+$ we say that d is the **greatest common divisor** of a and b, denoted by $\gcd(a, b)$ provided that:

i $d \mid a$ and $d \mid b$

ii if $c \mid a$ and $c \mid b$ then $c \leq d$

One way of finding the greatest common divisor of two positive integers is to first list all the divisors of each number, collect all the common divisors and then identify the greatest common divisor. For example:

Let $a = 12$ and $b = 30$. Let D_{12} and D_{30} be the sets of the divisors of 12 and 30 respectively. Then:

$D_{12} = \{1, 2, 3, 4, 6, 12\}$

$D_{30} = \{1, 2, 3, 5, 6, 10, 15, 30\}$

The common divisors are 1, 2, 3, 6 and so the greatest common divisor is 6.

Therefore $\gcd(12, 30) = 6$.

Moreover, we can use the **Euclidean algorithm** to find the greatest common divisor of two numbers a and b where $a, b \in \mathbb{Z}^+$, $b < a$. We divide the bigger of the two numbers, a, by the smaller one, b, to find the remainder r_1 where $r_1 < b$. Then we divide b by r_1 to obtain a remainder r_2. We then divide r_1 by r_2 to obtain a remainder r_3. We continue with this process until we reach $r_k = 0$. Then $r_{k-1} = \gcd(a, b)$.

The following example will illustrate the Euclidean algorithm.

Example 5

Find the greatest common divisor of 256 and 56.

$256 = 56 \times 4 + 32$	$a = 256$, $b = 56$ and $r_1 = 32$
$56 = 32 \times 1 + 24$	$r_2 = 24$
$32 = 24 \times 1 + 8$	$r_3 = 8$
$24 = 8 \times 3 + 0$	$r_4 = 0$
Therefore $\gcd(258, 56) = 8$	$r_4 = 0 \Rightarrow r_3 = \gcd(a, b)$

Theorem 3

Let $a, b \in \mathbb{Z}^+$, $a > b$, $d = \gcd(a, b)$. Then we can find $m, n \in \mathbb{Z}$ such that $d = ma + nb$.

Proof:

Using the Euclidean algorithm we have:

$a = b \times d_1 + r_1$
$b = r_1 \times d_2 + r_2$
$r_1 = r_2 \times d_3 + r_3$
$r_2 = r_3 \times d_4 + r_4$
.
.
$r_{k-3} = r_{k-2} \times d_{k-1} + r_{k-1}$
$r_{k-2} = r_{k-1} \times d_k + 0$

Working our way up this algorithm we obtain

$\gcd(a, b) = r_{k-1}$

$= r_{k-3} - r_{k-2} \times d_{k-1}$

$= r_{k-3} - (r_{k-4} - r_{k-3} \times d_{k-2}) \times d_{k-1}$ substituting for r_{k-2} from the previous line.

$= r_{k-3}(1 + d_{k-2}d_{k-1}) - r_{k-4} \times d_{k-1}$ rearranging and taking r_{k-3} as a common factor.

.
.

$= ma + nb$ We continue going backwards in this way until we obtain an expression in terms of a and b and since $r_i, d_i \in \mathbb{Z}$, m and n are integers. Q.E.D.

The next example illustrates how this theorem is used.

Example 6

Use the Euclidean algorithm to find the greatest common divisor of 28 and 36.
Hence find $m, n \in \mathbb{Z}$ such that $\gcd(28, 36) = 28m + 36n$.

$36 = 28 \times 1 + 8$ $28 = 8 \times 3 + 4$ $8 = 4 \times 2 + 0$ Therefore $\gcd(28, 36) = 4$.	*Use the Euclidean Algorithm.*
$4 = 28 - 8 \times 3$ $= 28 - 3(36 - 28 \times 1)$ $= 4 \times 28 - 3 \times 36$ Therefore $m = 4$ and $n = -3$.	*Work backwards using penultimate line of the algorithm, substituting for 8 from the previous line.* *Rearrange to obtain the required equation.*

In Example 7 you will see how this algorithm also works for larger numbers.

Example 7

Use the Euclidean algorithm to find the greatest common divisor of 721 and 448. Hence find $m, n \in \mathbb{Z}$ such that $\gcd(721, 448) = 721m + 448n$.	
$721 = 448 \times 1 + 273$ $448 = 273 \times 1 + 175$ $273 = 175 \times 1 + 98$ $175 = 98 \times 1 + 77$ $98 = 77 \times 1 + 21$ $77 = 21 \times 3 + 14$ $21 = 14 \times 1 + 7$ $14 = 7 \times 2 + 0$	*Use the Euclidean Algorithm.*
Therefore $\gcd(721, 448) = 7$. $\gcd(721, 448) = 7$ $= 21 - 14 \times 1$ $= 21 - (77 - 21 \times 3)$	*Using penultimate line of Algorithm, substituting for 14 from previous line.*
$= 4 \times 21 - 77$ $= 4(98 - 77) - 77$ $= 4 \times 98 - 5 \times 77$ $= 4 \times 98 - 5(175 - 98)$ $= 9 \times 98 - 5 \times 175$ $= 9(273 - 175) - 5 \times 175$ $= 9 \times 273 - 14 \times 175$ $= 9 \times 273 - 14(448 - 273)$ $= 23 \times 273 - 14 \times 448$ $= 23(721 - 448) - 14 \times 448$ $= 23 \times 721 - 37 \times 448$ Therefore $m = 23$ and $n = -37$.	*Rearrange.* *Substitute for 21 from previous line.* *Rearrange etc...*

Here is another example to help you understand how the algorithm works.

Example 8

Show that $\gcd(a - kb, b) = \gcd(a, b)$ where $a, b, k \in \mathbb{Z}^+$, $a > b$.	
Let $\gcd(a, b) = d$ and $\gcd(a - kb, b) = D$. $\gcd(a, b) = d \Rightarrow d \mid a$ and $d \mid b$ $d \mid a \Rightarrow a = md$ $d \mid b \Rightarrow b = nd \Rightarrow bk = knd$ $\therefore\ a - bk = (m - kn)d$ $\Rightarrow d \mid a - bk$	*Subtracting the two equations.*
So d is a common divisor of $a - kb$ and b. But since D is the $\gcd(a - kb, b)$ we can say that $d \le D$.	
Now $\gcd(a - kb, b) = D \Rightarrow D \mid a - kb$ and $D \mid b$ $D \mid a - kb \Rightarrow a - kb = pD$ $D \mid b \Rightarrow b = qD \Rightarrow kb = kqD$ $\therefore\ a = (p + kq)D$ $\Rightarrow D \mid a$ Therefore D is a common divisor of a and b. But since $\gcd(a, b) = d$ we can say that $D \le d$.	*Adding the two equations.*
Combining the two results we obtain $d \le D$ and $D \le d$ $\Rightarrow d = D \Rightarrow \gcd(a - kb, b) = \gcd(a, b)$.	

Exercise 1B

1 Find the gcd of each pair of numbers.

a 2406 and 654 **b** 728 and 548

c 1752 and 672 **d** 2595 and 1014

2 Using the results of question **1**, find integers m and n satisfying these equations.

a $2406m + 654n = \gcd(2406, 654)$

b $728m + 548n = \gcd(728, 548)$

c $1752m + 672n = \gcd(1752, 672)$

d $2595m + 1014n = \gcd(2595, 1014)$

EXAM-STYLE QUESTION

3 **a** Show that for $a, b \in \mathbb{Z}^+$ and $m, n \in \mathbb{Z}$, if $ma + nb = 1 \Rightarrow \gcd(a, b) = 1$.

b Hence show that if $\gcd(a, b) = 1$ and $\gcd(a, c) = 1$, then $\gcd(a, bc) = 1$.

Diophantus of Alexandria

DIOPHANTI
ALEXANDRINI
ARITHMETICORVM
LIBRI SEX,
ET DE NVMERIS MVLTANGVLIS
LIBER VNVS.
Nunc primum Græcè & Latinè editi, atque absolutissimis
Commentariis illustrati.
AVCTORE CLAVDIO GASPARE BACHETO
MEZIRIACO SEBVSIANO, V.C.
LVTETIAE PARISIORVM,
Sumptibus HIERONYMI DROVART, via Iacobæa,
sub Scuto Solari.
M. DC. XXI.
CVM PRIVILEGIO REGIS.

Diophantus is often referred to as "the father of algebra" due to his work on the solution of algebraic equations found in his work *Arithmetica*. There is very little known about his life; for example, it is not known exactly when he lived despite a number of speculations and estimations based on references to his writings. Diophantus dealt with positive rational solutions concerning linear and quadratic equations. These equations can be represented by $P(x, y, z\ldots) = 0$, e.g. $2x + 3y - 11 = 0$ or $5x^2 - 3y^2 - 4x + 7y - 13 = 0$

Linear Diophantine equations are sometimes referred to as first-order Diophantine equations.

The first example, $2x + 3y - 11 = 0$, is such an equation.
A Diophantine equation is one such equation for which only integer solutions are allowed.

Investigation – Diophantus Riddle

In this investigation you are asked to find a solution to a Diophantus Riddle which is presented in verse as follows:

Here lies Diophantus, the wonder behold
Through art algebraic, the stone tells how old
God gave him his boyhood one-sixth of his life,
One-twelfth more as youth while whiskers grew rife
And then yet one-seventh ere marriage begun;
In five years there came a bouncing new son.
Alas, this dear child of master and sage,
Attained only half of his father's age,
When chill fate took him. An event full of tears –
Heartbroken, his father lived just four more years.

1 Translate the poem into mathematical statements.

2 Let $n \in \mathbb{Z}^+$ be the age at which Diophantus died and let $m \in \mathbb{Z}^+$ be the number of years lived by his son. Use the statements in part **1** to form a pair of simultaneous equations.

3 Solve the equations to find how old Diophantus was when he died.

Linear Diophantine equations

Definition

A **linear Diophantine equation** in two variables is an equation of the form $ax + by = c$, where $a, b, c \in \mathbb{Z}$ and which has integer solutions x and y.

A simple linear equation in two variables $3x + 0y = 6$ is a linear Diophantine equation with infinitely many solutions $x = 2, y = n$, $n \in \mathbb{Z}$. Note that the solution of a Diophantine equation is made up of two parts.

? David Hilbert, a German mathematician from the University of Göttingen, presented 23 open mathematical questions at the International Congress of Mathematics in Sorbonne in 1900. Hilbert was convinced that all 23 problems would be solved. In fact, an engraving on his tombstone reads "We must know! We will know!" The problems on Hilbert's list have received remarkable attention since they were first presented, and four of them remain unsolved. The tenth problem in Hilbert's list of problems at the turn of the twentieth century is about finding an algorithm for determining solutions to general Diophantine equations. In 1970 Yuri Matiyasevich proved that it is not possible to find a general integer solution to all Diophantine equations. An algorithm does exist for the solution of linear Diophantine equations and you will be introduced to this algorithm in this section.

Theorem 4

A linear Diophantine equation $ax + by = c$, where $a, b, c \in \mathbb{Z}$, has integer solutions x and $y \in \mathbb{Z}$ if and only if $\gcd(a, b) \mid c$.

Proof:

$\Rightarrow$:

Let $x_1, y_1 \in \mathbb{Z}$ be solutions to the equation $\Rightarrow ax_1 + by_1 = c$

By definition, $\gcd(a, b) \mid a$ and $\gcd(a, b) \mid b$.

$\Rightarrow \gcd(a, b) \mid x_1a + y_1b \Rightarrow \gcd(a, b) \mid c$ (using Theorem 1)

$\Leftarrow$:

Let $\gcd(a, b) = d$

Then $\gcd(a, b) \mid c \Rightarrow d \mid c \Rightarrow c = md,\ m \in \mathbb{Z}$

But we know that there exist some $x_1, y_1 \in \mathbb{Z}$ such that $ax_1 + by_1 = d$ (using Theorem 3)

$\Rightarrow c = m(ax_1 + by_1) \Rightarrow c = a(mx_1) + b(my_1)$

Which means that the integers $x = mx_1$ and $y = my_1$ are solutions to the equation. Q.E.D.

Corollary

Given the linear Diophantine equation $ax + by = c$, where $a, b, c \in \mathbb{Z}$, if $\gcd(a, b) \nmid c$ there are no integer solutions to the equation.

Proof:

Suppose $\gcd(a, b) = d$ and $d \nmid c$.

Let $x_1, y_1 \in \mathbb{Z}$ be solutions to the equation.

Then since $d \mid a$ and $d \mid b$ it follows by theorem 3 that $d \mid ax_1 + by_1 = c$ which is a contradiction.

Hence, if $\gcd(a, b) \nmid c$ then there are no integer solutions to the equation. Q.E.D.

Before we move on to find an algorithm for the general solution of a linear diophantine equation in two variables we need to prove the following theorem about greatest common divisors.

Theorem 5

If $\gcd(a, b) = d$, then $\gcd\left(\frac{a}{d}, \frac{b}{d}\right) = 1$.

Proof:

Let $\gcd(a, b) = d \Rightarrow a = md$ and $b = nd$, and m, n are relatively prime.

If m, n are not relatively prime then d is not the greatest common divisor of a and b.

$\gcd\left(\frac{a}{d}, \frac{b}{d}\right) = \gcd\left(\frac{md}{d}, \frac{nd}{d}\right) = \gcd(m, n) = 1$ Q.E.D.

Theorem 6

Given $ax + by = c$, where $\gcd(a, b) \mid c$, and x_0, y_0 is a particular solution then $\left\{x_0 + \left(\frac{b}{d}\right)k, y_0 - \left(\frac{a}{d}\right)k, k \in \mathbb{Z}\right\}$ is a complete set of solutions of the given Diophantine equation.

Proof:

The proof is split into two parts. First we must show that any pair of the form $x_0 + \left(\frac{b}{d}\right)k, y_0 - \left(\frac{a}{d}\right)k, k \in \mathbb{Z}$ is a solution.

Then we must show that $\left\{x_0 + \left(\frac{b}{d}\right)k, y_0 - \left(\frac{a}{d}\right)k \middle| k \in \mathbb{Z}\right\}$ gives the whole solution set.

In order to show that $x_0 + \left(\frac{b}{d}\right)k, y_0 - \left(\frac{a}{d}\right)k, k \in \mathbb{Z}$ is a solution we need to substitute for x and y in the LHS of the given equation.

In other words, $a\left(x_0+\left(\frac{b}{d}\right)k\right)+b\left(y_0-\left(\frac{a}{d}\right)k\right)$

$$= ax_0 + a\left(\frac{b}{d}\right)k + by_0 - b\left(\frac{a}{d}\right)k$$

$$= ax_0 + by_0 + a\left(\frac{b}{d}\right)k - b\left(\frac{a}{d}\right)k$$

$$= c + 0 = c$$

Since x_0, y_0 are particular solutions it follows that $ax_0 + by_0 = c$.

Therefore the pair $x_0+\left(\frac{b}{d}\right)k,\ y_0-\left(\frac{a}{d}\right)k$ does give a solution to the given equation.

To show that the solution set is complete let us take a general solution x, y.

Then $ax_0 + by_0 = c = ax + by$

$\Rightarrow a(x - x_0) = -b(y - y_0)$

$\Rightarrow \frac{a}{d}(x - x_0) = -\frac{b}{d}(y - y_0)$

$\frac{a}{d}|LHS \Rightarrow \frac{a}{d}\,\Big|\, -\frac{b}{d}(y - y_0)$

But by Theorem 5 $\gcd\left(\frac{a}{d}, \frac{b}{d}\right) = 1 \Rightarrow \frac{a}{d} \nmid \frac{b}{d}$

Therefore $\frac{a}{d}|-(y - y_0) \Rightarrow -(y - y_0) = \frac{a}{d}k$

$$\Rightarrow y = y_0 - \left(\frac{a}{d}\right)k$$

Now substitute for $y - y_0$ into $\frac{a}{d}(x - x_0) = -\frac{b}{d}(y - y_0)$

$\Rightarrow \frac{a}{d}(x - x_0) = -\frac{b}{d}\left(-\frac{a}{d}k\right)$

$\Rightarrow x - x_0 = -\frac{b}{a}\left(-\frac{a}{d}k\right) \Rightarrow x = x_0 + \left(\frac{b}{d}\right)k$

Hence we have shown that $x_0+\left(\frac{b}{d}\right)k,\ y_0-\left(\frac{a}{d}\right)k$ where $k \in \mathbb{Z}$ give a complete infinite solution set of the linear Diophantine equation $ax + by = c$ where $\gcd(a, b)\,|\,c$. Q.E.D.

Although it is sometimes easy to find the gdc of two numbers by finding their factors, we use the Euclidean algorithm for finding the particular solution of a linear Diophantine equation.

The following examples show how you can use this result to find the general solutions of linear Diophantine equations.

Example 9

Determine which of the following Diophantine equations have solutions.
If an equation has a solution, find **i** a particular solution, and **ii** a general solution.

a $3x + 4y = 1$ **b** $2x + 5y = 12$

c $3x + 6y = 7$ **d** $15x + 12y = 105$

a gcd(3, 4) = 1	*Since the RHS is 1 and gcd(3, 4) = 1.*
The equation has a solution.	
$1 = 4 - 1 \times 3 = (-1) \times 3 + (1) \times 4$	
Particular solution is $x_0 = -1$, $y_0 = 1$.	*Compare with given equation to find particular solution.*
The complete solution is therefore $x = -1 + 4k$, $y = 1 - 3k$, $k \in \mathbb{Z}$.	*Applying the formulae* $x = x_0 + \left(\frac{b}{d}\right)k,\ y = y_0 - \left(\frac{a}{d}\right)k$
b	
Method I	
gcd(5, 2) = 1	*Since both 5 and 2 are prime numbers.*
This results in	
$1 = 5 - 2 \times 2 = (-2) \times 2 + 1 \times 5$	*The equation has a solution since gcd(5, 2) = 1.*
So, $12 = (-24) \times 2 + 12 \times 5$,	*Multiply by 12.*
giving the particular solution: $x_0 = -24$, $y_0 = 12$.	*Compare with given equation to find particular solution.*
The complete solution is therefore $x = -24 + 5k$, $y = 12 - 2k$, $k \in \mathbb{Z}$.	*Applying the formulae* $x = x_0 + \left(\frac{b}{d}\right)k,\ y = y_0 - \left(\frac{a}{d}\right)k$
Method II	
By inspection we note that a particular solution is $x_0 = 1$, $y_0 = 2$.	*When numbers are small it is easier to find a particular solution by inspection.*
The complete solution is therefore $x = 1 + 5k$, $y = 2 - 2k$, $k \in \mathbb{Z}$.	*You can check that this general solution is equivalent to the general solution obtained using the algorithm in Method I.*
c $3x + 6y = 7$	*7 is not a multiple of 3.*
gcd(3, 6) = 3 and the equation does not have a solution.	
d $15x + 12y = 105$	*Find the gcd(15, 12) using the Euclidean algorithm.*
$15 = 12 \times 1 + 3$	
$12 = 3 \times 1 + 0$	
gcd(15, 12) = 3 and $3 \mid 105$, so a solution exists.	*Check whether 105 is a multiple of gcd(15, 12).*
$3 = 15 \times 1 + 12 \times (-1)$	*Work backwards and multiply by 35.*
$105 = 15 \times 35 + 12 \times (-35)$, giving a particular solution $x_0 = 35$, $y_0 = -35$	*Compare with the given equation.*
The complete solution is therefore: $x = 35 + 4k$, $y = -35 - 5k$, $k \in \mathbb{Z}$.	*Applying the formulae* $x = x_0 + \left(\frac{b}{d}\right)k,\ y = y_0 - \left(\frac{a}{d}\right)k$

Diophantine equations are also used to find solutions of some real-life problems. The next example illustrates how to do this.

Example 10

As Beppe prepares for his Mathematics HL exam he finds that he can solve a short response question in 5 minutes but it takes him 18 minutes to solve a long response question. What combination of complete questions can he answer if he works continuously for 96 minutes without losing any time?

$5x + 18y = 96$	*Write the information algebraically.*
$18 = 3 \times 5 + 3$ $5 = 1 \times 3 + 2$ $3 = 1 \times 2 + 1$ Since gcd(18, 5) = 1 the equation has a solution.	*Find gcd(18, 5).*
$1 = 3 - 2$ $= 3 - (5 - 1 \times 3)$ $= 2 \times 3 - 5$ $= 2 \times (18 - 3 \times 5) - 5$ $= 5 \times (-7) + 2 \times 18$	*Working backwards.*
$96 = 5 \times (-672) + 18 \times 192$ Particular solution is $x_0 = -672, y_0 = 192$	*Multiply by 96.*
General solution: $x = -672 + 18k, y = 192 - 5k, k \in \mathbb{Z}$.	*Applying the formulae* $x = x_0 + \left(\frac{b}{d}\right)k,\ y = y_0 - \left(\frac{a}{d}\right)k$
Now we must apply the conditions defined by the question: $0 \le -672 + 18k < 20 \Rightarrow 37 < k \le 39$ $0 \le 192 - 5k < 6 \Rightarrow 38 \ge k > 37$ Therefore $k = 38$ $\Rightarrow x = -672 + 18 \times 38 = 12$ $y = 192 - 5 \times 38 = 2$ So Beppe can solve 12 short response and 2 long response questions without wasting any time.	*Beppe might work only on long response questions, or only on short response questions. In 96 minutes he would respond to at most 19 full short questions or at least 5 full long response questions. Solving these inequalities and substituting to find x and y we obtain the combination of possible complete questions that Beppe can answer in 96 minutes.*

Exercise 1C

In questions 1 to 10, determine which of the linear Diophantine equations have a solution. If a solution exists find:

i a particular solution

ii a general solution.

1 $5x + 3y = 1$

2 $184x + 76y = 1$

3 $3x + 8y = 15$

4 $6x + 8y = 11$

5 $14x + 21y = 28$

6 $90x + 15y = 128$

7 $90x + 15y = 135$

8 $125x + 60y = 200$

9 $1769x + 238y = 3000$

10 $2311x + 1137y = 1543$

11 Let $c \in \mathbb{Z}^+$, $10 < c < 20$. Determine the values of c for which the Diophantine equation $84x + 990y = c$ has no solutions. Find the general solutions for the other values of c.

12 Gino collected €100 in funds towards an animal shelter where he carries out a CAS project. He decided to spend this money on treats for dogs and cats. A bag of dog treats costs €16 and a bag of cat treats costs €12. What combinations of dog treats and cat treats can he buy if he spends all the money collected?

Prime Numbers

Definition

A positive integer p greater than 1 is said to be **prime** if the only positive factors of p are 1 and p itself. A positive integer greater than 1 which is not prime is said to be **composite**.

The number 1 is not a prime number because it is a very special number forming the building block of all positive integers. It is the only number that leaves a number unchanged upon multiplication by it. It is the only positive integer with only one positive divisor and it is the only number that remains unchanged when raised to any power.

The integer 13 is prime since it can be divided by only 1 and 13. 6 is not a prime number since $2|6$ and $3|6$, therefore 2 and 3 are factors of 6.

The integer 2 is the smallest, and the only even, prime number.

The sieve of Eratosthenes is an ancient iterative algorithm for finding prime numbers. To this day it is one of the most efficient ways for finding smaller prime numbers. Although a number of interesting patterns emerge when shading in the non-primes, to this date no pattern has been found for the primes themselves.

	2	3	4	5	6	7	8	9	10
11	12	13	14	15	16	17	18	19	20
21	22	23	24	25	26	27	28	29	30
31	32	33	34	35	36	37	38	39	40
41	42	43	44	45	46	47	48	49	50
51	52	53	54	55	56	57	58	59	60
61	62	63	64	65	66	67	68	69	70
71	72	73	74	75	76	77	78	79	80
81	82	83	84	85	86	87	88	89	90
91	92	93	94	95	96	97	98	99	100

Theorem 7

If n is a composite positive integer, then n has a factor less than or equal to $\sqrt{n}$.

Proof:

Given that n is composite there must be $a, b \in \mathbb{Z}^+$, such that $1 < a \leq b < n$, which are factors of n. In other words,

$a \leq b \Rightarrow a^2 \leq ab = n \Rightarrow a \leq \sqrt{n}$ Q.E.D.

This theorem provides us with another way of finding prime numbers as demostrated by the next example.

Example 11

Determine which of these are prime numbers.

a 101 **b** 247 **c** 163

a $\sqrt{101} < 11$ 101 is prime since none of the numbers 2, 3, 5, 7 are factors of 101.	*The prime numbers less than 11 are 2, 3, 5, 7.*
b $\sqrt{247} < 16$ $13 \mid 247$ since $13 \times 19 = 247$ Therefore 247 is not a prime number.	*Check for divisibility by the prime numbers less than 16.*
c $\sqrt{163} < 13$ None of the numbers 2, 3, 5, 7, 11 are factors of 163. Therefore 163 is a prime number.	*Check for divisibility by the prime numbers less than 13.*

In Hilbert's list of problems at the turn of the 20th century, problem 8 is about prime numbers and their distribution, which gave birth to the Riemann Hypothesis, named after another famous German mathematician, Bernhard Riemann. An intriguing question that to this day remains unanswered asks whether we can find a formula that generates prime numbers. Riemann attacked this problem from a completely new perspective and began to find some patterns in the elusive and chaotic distribution of the primes. His prediction about the distribution of the primes is known as the Riemann Hypothesis and to this day it has not been proved. In fact, a solution to this hypothesis would have great implications on the modern world. Although one might think that the distribution of the primes is simply a mathematical challenge for mathematicians, prime numbers play a central role in e-commerce and trading on the internet by providing a secure system known as RSA. The security of this system depends on the elusive nature of prime numbers.

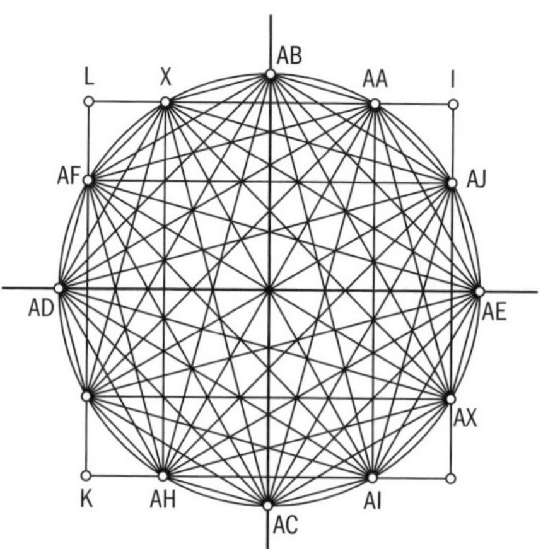

The Clay Mathematics Institute is offering $1 million to any individual or group that can solve the Riemann Hypothesis. The diagram here is a suggested proof of this hypothesis, found on: http://blog.fora.tv/2014/05/beauty-in-numbers-solving-the-unsolvable-riemann-hypothesis

Definition

Two integers a and b are said to be **relatively prime** or **co-prime** if and only if $\gcd(a, b) = 1$.

Example 12

Use the Euclidean algorithm to show that 17 and 22 are relatively prime.
Hence find integers m and n such that $17m + 22n = 1$.

$22 = 17 \times 1 + 5$ $17 = 5 \times 3 + 2$ $5 = 2 \times 2 + 1$ $2 = 2 \times 1 + 0$ Therefore $\gcd(17, 22) = 1$.	*Use the Euclidean algorithm to find the gcd(22, 17).*
Using the above: $1 = 5 - 2 \times 2$ $= 5 - 2(17 - 5 \times 3)$ $= 5 \times 7 - 17 \times 2$ $= 7(22 - 17) - 17 \times 2$ $= 7 \times 22 - 9 \times 17$ Therefore $m = -9$ and $n = 7$.	*Work backwards from the Euclidean algorithm to find m and n.*

The positive integers can be categorized into three distinct sets:

- Prime numbers, which are those numbers that can be divided by only 1 or themselves.
- Composite numbers, which are numbers that can be written as a product of smaller prime numbers.
- The number 1 which is neither prime nor composite.

The following theorem was proved by the Greek mathematician Euclid. It is a very simple but elegant proof by contradiction.

Theorem 8 (Euclid's statement):

There are infinitely many prime numbers.

Proof:

Assume that there is a finite number of prime numbers. We can then list them as follows: $p_1, p_2, p_3, \ldots, p_n$ where p_n is the largest prime number.

Now construct a new integer $m = p_1 \times p_2 \times p_3 \times \ldots \times p_n + 1$.

This new number cannot be prime because $m > p_n$ and we said that p_n is the largest prime number.

If m is divided by any of the prime numbers $p_1, p_2, p_3, \ldots, p_n$ it will leave a remainder of 1. So either m is prime or it is divisible by another prime number which is not one of the list $p_1, p_2, p_3, \ldots, p_n$. Hence by contradiction there must be an infinite number of primes. Q.E.D.

Investigation - Mersenne primes and perfect numbers

Mersenne primes are prime numbers that can be written in the form $2^n - 1$. They are named after the 17th century French mathematician Marin Mersenne. Not all numbers of this form produce prime numbers and the smallest non-prime Mersenne number is $2^{11} - 1$.

A perfect number is a positive integer that is equal to the sum of its positive divisors, excluding the number itself. The smallest perfect number is $6 = 1 + 2 + 3$.

Copy and complete the following table:

n	$\sum_{k=0}^{n} 2^k$	$2^{n+1} - 1$	Prime	$2^n \times \sum_{k=0}^{n} 2^k$	Perfect Number
1	$1 + 2 = 3$	$2^2 - 1 = 3$	Yes	$2 \times 3 = 6$	$1 + 2 + 3 = 6$
2	$1 + 2 + 4 = 7$		Yes	$4 \times 7 = 28$	$1 + 2 + 4 + 7 + 14 = 28$
3	$1 + 2 + 4 + 8 = 15$		No		
4					
5					

Use your results to make conjectures about $\sum_{k=0}^{n} 2^k$.

Use mathematical induction to prove your conjecture.

Make a conjecture connecting Mersenne primes and perfect numbers.

Exercise 1D

1 Explaining your method fully, determine whether or not 493 is a prime number.

2 Write 19152 as a product of primes.

3 Use the Euclidean algorithm to show that 250 and 111 are relatively prime. Hence find integers m and n such that $250m + 111n = 1$.

4 If a and b are relatively prime and $a > b$, prove that $\gcd(a - b, a + b)$ is either 1 or 2.

5 Use the Euclidean algorithm to show that for $n \in \mathbb{Z}^+$, the positive integers $5n + 3$ and $7n + 4$ are relatively prime.

1.3 Strong mathematical induction

Before we continue our work with prime numbers we need to look at a new method of proof. In the core syllabus, you were introduced to proving statements using mathematical induction, sometimes referred to as **weak** mathematical induction. We will now take this a step further by looking at **strong** mathematical induction. Essentially the difference between the two methods is in the inductive step.

In weak induction, the format of a proof is as follows:

- State the claim that is being asserted P_n.
- Prove that the claim is true for an initial value of n, e.g. when $n = 1$ (n does not necessarily have to be 1).
- Assume that the claim is true for **some** $n = k$, $k \geq 1$. (This is the inductive step.)
- Show that if this assumption is true, the statement is also true for $n = k + 1$.
- Make a final statement to show that through the inductive process the statement is true for all values of n.

In strong induction the same format is used, however the inductive step changes to the following:

- Assume that the claim is true **for all** i, such that $1 \leq i \leq k$.

So the difference is in the assumption that you make in the inductive step. The terms "weak" and "strong" can be confusing. The names do not mean that all the proofs done using weak induction are weaker proofs. It means that we proved the claim going one step at a time starting from the initial step. In strong induction the assumptions in the inductive step appear to be more demanding than the assumption in the weak case. Often the base case for strong induction involves more than just $n = 1$.

The next example illustrates how strong induction can be used to prove that every positive integer greater than 1 can be written in binary form.

Example 13

Show that every positive integer n can be expressed in the form: $n = a_r 2^r + a_{r-1}2^{r-1} + a_{r-2}2^{r-2} + \ldots + a_0 2^0$	
Proof:	
$P_n : n = a_r 2^r + a_{r-1}2^{r-1} + a_{r-2}2^{r-2} + \ldots + a_0 2^0$	*State the claim you want to prove.*
When $n = 1$, $1 = 2^0$.	*Prove that the claim is true for the first positive integer.*
Assume that P_n is true for all $1 \le i \le k$. When $n = k + 1$ then it can be even or it can be odd.	*Make the assumption (inductive step).*
If $k + 1$ is even, then $\frac{k+1}{2}$ is an integer which is less than k and by the inductive step it may be written as $\frac{k+1}{2} = a_r 2^r + a_{r-1}2^{r-1} + a_{r-2}2^{r-2} + \ldots + a_0 2^0$ $\Rightarrow k + 1 = a_r 2^{r+1} + a_{r-1}2^r + a_{r-2}2^{r-1} + \ldots + a_0 2$ If $k + 1$ is odd then $\frac{(k+1)-1}{2}$ is an integer which is less than k and by the inductive step it may be written as $\frac{k}{2} = a_r 2^r + a_{r-1}2^{r-1} + a_{r-2}2^{r-2} + \ldots + a_0 2^0$ $\Rightarrow k = a_r 2^{r+1} + a_{r-1}2^r + a_{r-2}2^{r-1} + \ldots + a_0 2$ $\Rightarrow k + 1 = a_r 2^{r+1} + a_{r-1}2^r + a_{r-2}2^{r-1} + \ldots + a_0 2 + 1$ i.e. $k + 1 = a_r 2^{r+1} + a_{r-1}2^r + a_{r-2}2^{r-1} + \ldots + a_0 2 + 2^0$	*Prove that the statement is true for n = k.*
Since P_1 is true and it was shown that given P_i is true for all $1 \le i \le k$, P_{k+1} is also true, it follows by the principle of strong induction that P_n is true for all $n \ge 1$, $n \in \mathbb{Z}^+$.	

The next example shows the proof of a statement that is true for all integers greater than 12, which is not intuitively obvious.

Example 14

Show that $n = 4p + 5q$ for all $p, q \in \mathbb{Z}^+$, $n \geq 12$, $n \in \mathbb{Z}$.	
Proof: $P_n: n = 4p + 5q, p, q \in \mathbb{Z}^+, n \geq 12, n \in \mathbb{Z}$	*State the claim that you want to prove.*
$12 = 3 \times 4 + 0 \times 5$ $13 = 2 \times 4 + 1 \times 5$ $14 = 1 \times 4 + 2 \times 5$ $15 = 0 \times 4 + 3 \times 5$ $16 = 4 \times 4 + 0 \times 5$	*The initial step requires that we prove the result for five consecutive integers greater than or equal to 12.*
Assume that P_i is true for all $15 \leq i \leq k$. When $n = k + 1$, $k + 1 = (k - 4) + 5$ $= 4p + 5q + 5$ $= 4p + 5(q + 1)$	*Inductive step.* *Rewrite $k + 1$ in terms of an integer less than k, e.g. $(k - 4)$ which can be written as $4p + 5q$.*
Since $P_{12}, P_{13}, P_{14}, P_{15}, P_{16}$ are true, and it was shown that given P_i is true for all $12 \leq i \leq k$, P_{k+1} is also true, it follows by the principle of strong induction that P_n is true for all $n \geq 12$, $n \in \mathbb{Z}^+$.	

Strong induction is used for proving recursive relations and inequalities.
The following example uses strong induction to prove an inequality.

Example 15

Show that if $u_n = u_{n-1} + u_{n-2}$ with $u_1 = 1$ and $u_2 = 2$, then for all $n \in \mathbb{Z}^+$ $u_n < 2^n$.	
Proof: $P_n: u_n < 2^n$ for all $n \in \mathbb{Z}^+$	*State the claim you want to prove.*
$u_1 = 1 < 2^1$ $u_2 = 2 < 4 = 2^2$ Therefore P_1 and P_2 are true. Assume that P_i is true for all $1 \leq i \leq k$. When $n = k + 1$ $u_{k+1} = u_k + u_{k-1}$ $< 2^k + 2^{k-1}$ $< 2^k + 2^k$ $\Rightarrow u_{k+1} < 2 \times 2^k = 2^{k+1}$	*Prove initial conditions hold.* *Inductive step.* In chapter 3 you will be studying recurrence relations in detail and you will encounter more proofs using strong induction.
Since P_1, P_2 are true, and it was shown that given P_i is true for all $1 \leq i \leq k$ then P_{k+1} is also true, it follows by the principle of strong induction that P_n is true for all $n \in \mathbb{Z}^+$.	

Exercise 1E

1 Show that $n = 3p + 8q$ for all $p, q \in \mathbb{Z}^+$, $n \geq 14$, $n \in \mathbb{Z}$.

2 Show that $n = 3x + 5y$ for all $x, y \in \mathbb{Z}^+$, $n > 7$, $n \in \mathbb{Z}^+$.

3 Prove that $2! \times 4! \times 6! \times \ldots \times (2n)! \geq ((n + 1)!)^n$

4 **a** Prove that $n^2 \geq 2n + 1$ for all $n \geq 3$, $n \in \mathbb{Z}^+$.

b Hence prove that $2^n \geq n^2$ for all $n \geq 4$, $n \in \mathbb{Z}^+$.

5 Consider the sequence defined by $T_1 = T_2 = 2$ and $T_n = T_{n-1} + T_{n-2}$ for $n \geq 3$. Prove that $T_n \leq 2^n$ for all $n \in \mathbb{Z}^+$.

1.4 The Fundamental Theorem of Arithmetic and least common multiples

In this section you will see why prime numbers are sometimes called the building blocks of positive integers. Having defined prime numbers and composite numbers we shall start by proving a very important result in the next theorem.

It is important to remember that this theorem only holds for a prime number p. If p is replaced by any positive integer the theorem does not hold. For example $6|30$ and $30 = 2 \times 15$ but $6 \nmid 2$ and $6 \nmid 15$.

Theorem 9

If p is a prime number and $a_1, a_2, a_3, a_4, \ldots, a_n \in \mathbb{Z}^+$ such that $p|(a_1 \times a_2 \times a_3 \times a_4 \times \ldots \times a_n)$, then $p|a_i$ for some a_i where $1 \leq i \leq n$.

Proof:

Using the principle of mathematical induction:

P_n: $p|(a_1 \times a_2 \times a_3 \times a_4 \times \ldots \times a_n) \Rightarrow p|a_i$ for some a_i where $1 \leq i \leq n$.

When $n = 1$ it becomes obvious from the statement that $p|a_1$.

When $n = 2$ we are given that $p|a_1a_2$.

The result for $n = 2$ is referred to as Euclid's Lemma.

Suppose that p does not divide a_1, then since p is a prime number we know that p and a_1 are relatively prime, i.e. $\gcd(p, a_1) = 1$.

Using Theorem 3, $\gcd(p, a_1) = 1 \Rightarrow 1 = mp + na_1$ where $m, n \in \mathbb{Z}$.

$$\begin{aligned} \text{Now } a_2 &= 1 \times a_2 \\ &= (mp + na_1) \times a_2 \\ &= mpa_2 + na_1a_2 \qquad (\text{since } p|a_1a_2 \Rightarrow a_1a_2 = kp,\ k \in \mathbb{Z}) \\ &= p(ma_2 + k) \end{aligned}$$

Therefore $p|a_2$.

Assume that $p|(a_1 \times a_2 \times a_3 \times a_4 \times \ldots \times a_n) \Rightarrow p|a_i$ for some a_i where $1 \leq i \leq k$.

When $n = k + 1$ we have $p|a_1a_2a_3a_4\ldots a_ka_{k+1}$.

Now let $a_1a_2a_3a_4...a_k = A$.

Then we have $p \mid Aa_{k+1}$.

If $p \mid A$ then by assumption $p \mid a_i$ for some a_i where $1 \le i \le k$.

If p does not divide A then by result in the basic step for $n = 2$, $p \mid a_{k+1}$.

Therefore we can say that $p \mid a_i$ for some a_i where $1 \le i \le k + 1$ and P_{k+1} is true.

Since we proved that P_1 and P_2 are true and we showed that if P_k is true then P_{k+1} is also true, it follows by the principle of mathematical induction that P_n is true for all $n \in \mathbb{Z}^+$.   Q.E.D.

Theorem 10: The Fundamental Theorem of Arithmetic

Every positive integer n greater than 1 can be written **uniquely** as a product of primes written in ascending order, i.e. $n \in \mathbb{Z}^+, n > 1 \Rightarrow n = p_1^{\alpha_1} p_2^{\alpha_2} p_3^{\alpha_3} ... p_i^{\alpha_i} ... p_m^{\alpha_m}$, where p_i is a prime number for all $1 \le i \le m$, $p_1 < p_2 < p_3 < ... < p_m$ and $\alpha_i \in \mathbb{N}$.

Proof:

First we need to prove the result that any positive integer can be factorized into prime numbers. We shall use strong induction to prove this.

$P_n : n = p_1^{\alpha_1} p_2^{\alpha_2} p_3^{\alpha_3} ... p_i^{\alpha_i} ... p_m^{\alpha_m}$, where p_i is prime, $n \in \mathbb{Z}^+$, $n > 1$.

2 is a prime number therefore it is a prime factorization of itself.

Therefore P_2 is true.

3 is a prime number therefore it is a prime factorization of itself.

Therefore P_3 is true.

4 is not a prime number, but $4 = 2^2$ which is a prime factorization.

Therefore P_4 is true.

Assume that for all $i \in \mathbb{Z}^+$, $2 \le i \le k$ we can write

$i = p_1^{\beta_1} p_2^{\beta_2} p_3^{\beta_3} ... p_i^{\beta_i} ... p_m^{\beta_m}$ where p_j is prime for all j.

If $k + 1$ is a prime number then it is a prime factorization of itself.

If $k + 1$ is not a prime number then it is composite, and there must be two integers $x, y \in \mathbb{Z}^+$, $1 < x < k + 1$ and $1 < y < k + 1$ such that $k + 1 = xy$.

But using the inductive step on x and y we can write $x = p_1^{\alpha_1} p_2^{\alpha_2} p_3^{\alpha_3} \dots p_n^{\alpha_n}$ and $y = p_1^{\beta_1} p_2^{\beta_2} p_3^{\beta_3} \dots p_i^{\beta_i} \dots p_n^{\beta_n}$.

Which means that $k + 1 = xy = p_1^{\alpha_1+\beta_1} p_2^{\alpha_2+\beta_2} p_3^{\alpha_3+\beta_3} \dots p_n^{\alpha_n+\beta_n}$.

Since $\alpha_i, \beta_i \in \mathbb{N}$ it follows that $\alpha_i + \beta_i \in \mathbb{N}$ for all $1 \le i \le n$.

Therefore P_{k+1} is true.

Since P_2, P_3 and P_4 are true and it was shown that given P_i is true for all $2 \le i \le k$, P_{k+1} is also true, it follows by the principle of strong induction that P_n is true for all $n \in \mathbb{Z}^+$, $n > 1$.

Now we need to prove uniqueness.

Assume that there is one number N for which the theorem is false, i.e. $N = p_1^{\alpha_1} p_2^{\alpha_2} p_3^{\alpha_3} \ldots p_i^{\alpha_i} \ldots p_n^{\alpha_n}$ and $N = p_1^{\beta_1} p_2^{\beta_2} p_3^{\beta_3} \ldots p_i^{\beta_i} \ldots p_n^{\beta_n}$ where p_i's are prime numbers not necessarily distinct and $\alpha_i, \beta_i \in \mathbb{N}$.

$$p_1^{\alpha_1} \times p_2^{\alpha_2} \times \ldots \times p_n^{\alpha_n} = p_1^{\beta_1} \times p_2^{\beta_2} \times \ldots \times p_n^{\beta n}$$

$$\Rightarrow p_1^{\alpha_1} \times p_2^{\alpha_2} \times \ldots \times p_n^{\alpha_n} \times (p_1^{-\beta_1} \times p_2^{-\beta_2} \times \ldots p_n^{-\beta_n}) = p_1^{\beta_1} \times p_2^{\beta_2} \times \ldots \times p_n^{\beta_n} \times (p_1^{-\beta_1} \times p_2^{-\beta_2} \times \ldots p_n^{-\beta_n})$$

$$\Rightarrow p_1^{\alpha_1} \times p_1^{-\beta_1} \times p_2^{\alpha_2} \times p_2^{-\beta_2} \times \ldots \times p_n^{\alpha_n} \times p_n^{-\beta_n} = p_1^{\beta_1} \times p_1^{-\beta_1} \times p_2^{\beta_2} \times p_2^{-\beta_2} \times \ldots \times p_n^{\beta_n} \times p_n^{-\beta_n}$$

$$\Rightarrow p_1^{\alpha_1-\beta_1} \times p_2^{\alpha_2-\beta_2} \times \ldots \times p_n^{\alpha_n-\beta_n} = p_1^{\beta_1-\beta_1} \times p_2^{\beta_2-\beta_2} \times \ldots \times p_n^{\beta_n-\beta_n} = 1$$

$$\Rightarrow p_1^{\alpha_1-\beta_1} \times p_2^{\alpha_2-\beta_2} \times \ldots \times p_n^{\alpha_n-\beta_n} = 1$$

Therefore, for every $1 \le i \le n \Rightarrow p_i^{\alpha_i-\beta_i} = 1 \Rightarrow \alpha_i - \beta_i = 0 \Rightarrow \alpha_i = \beta_i$. Q.E.D.

An alternative direct proof is as follows:

Let $N = p_1 \times p_2 \times \ldots \times p_n = q_1 \times q_2 \times \ldots \times q_m$ where p_i, q_i are prime numbers and $n < m$.

Since $p_1 \mid N$ and $N = q_1 q_2 q_3 \ldots q_m$ it follows from theorem 9 that p_1 must divide some q_i. But q_i is prime so $p_1 = q_i$ and thus we can cancel p_1 and q_i from the equation. Now we can repeat the same procedure and cancel p_2 and q_j, and so on, until all the prime numbers on the LHS of the equation are exhausted and we are left with $1 = \prod_{k=1}^{m-n} q_k$ which is a contradiction since a product of prime numbers can never be equal to 1. Thus all the factors on each side of the equation must be the same.

Definition

The **least common multiple** of two positive integers a and b, denoted by lcm(a, b) is the smallest positive integer that is divisible by both a and b. Given $a, b \in \mathbb{Z}^+$, $\text{lcm}(a, b) = m \Rightarrow a \mid m$ and $b \mid m$ and if there is another $n \in \mathbb{Z}^+$ such that $a \mid n$ and $b \mid n$ then $m \le n$.

The Fundamental Theorem of Arithmetic (FTA) provides us with a system of finding the least common multiple and the greatest common divisor of two positive integers as follows.

Let $x = p_1^{\alpha_1} p_2^{\alpha_2} p_3^{\alpha_3} \ldots p_k^{\alpha_k}$ and $y = p_1^{\beta_1} p_2^{\beta_2} p_3^{\beta_3} \ldots p_k^{\beta_k}$

Then the $\text{lcm}(a, b) = p_1^{\max(\alpha_1,\beta_1)} p_2^{\max(\alpha_2,\beta_2)} p_3^{\max(\alpha_3,\beta_3)} \ldots p_k^{\max(\alpha_k,\beta_k)}$

and consequently $\gcd(a, b) = p_1^{\min(\alpha_1,\beta_1)} p_2^{\min(\alpha_2,\beta_2)} p_3^{\min(\alpha_3,\beta_3)} \ldots p_k^{\min(\alpha_k,\beta_k)}$

For example, if we want to find lcm(28, 72) and gcd(28, 72) we first write the prime factorization of 28 and 72.

$28 = 2^2 \times 3^0 \times 7^1$

$72 = 2^3 \times 3^2 \times 7^0$

So, $\text{lcm}(28, 72) = 2^3 \times 3^2 \times 7^1 = 504$

and $\text{gcd}(28, 72) = 2^2 \times 3^0 \times 7^0 = 4$.

The solutions can be also found on a GDC:

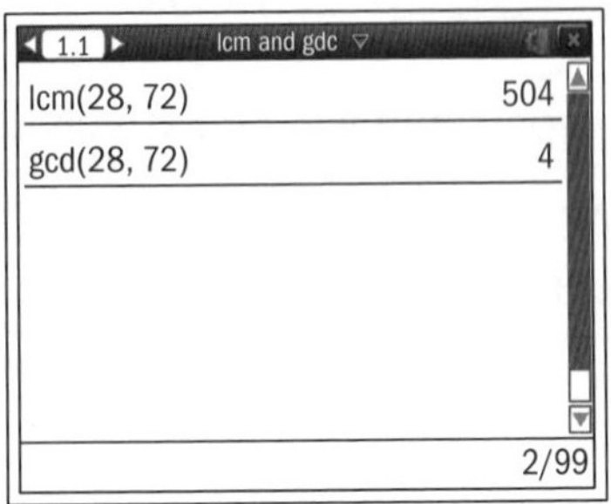

Example 16

Given that $\gcd(x, y) = m$ and $\text{lcm}(x, y) = n$, show that $xy = mn$.

$x = p_1^{\alpha_1} p_2^{\alpha_2} p_3^{\alpha_3} \ldots p_k^{\alpha_k}$ $y = p_1^{\beta_1} p_2^{\beta_2} p_3^{\beta_3} \ldots p_k^{\beta_k}$	*Using FTA.*
Then $\text{lcm}(x, y) = p_1^{\max(\alpha_1,\beta_1)} p_2^{\max(\alpha_2,\beta_2)} p_3^{\max(\alpha_3,\beta_3)} \ldots p_k^{\max(\alpha_k,\beta_k)} = n$	*By definition of least common multiple.*
$\gcd(x, y) = p_1^{\min(\alpha_1,\beta_1)} p_2^{\min(\alpha_2,\beta_2)} p_3^{\min(\alpha_3,\beta_3)} \ldots p_k^{\min(\alpha_k,\beta_k)} = m$ $\Rightarrow mn = p_1^{\max(\alpha_1,\beta_1)} p_2^{\max(\alpha_2,\beta_2)} p_3^{\max(\alpha_3,\beta_3)} \ldots p_k^{\max(\alpha_k,\beta_k)}$ $\times p_1^{\min(\alpha_1,\beta_1)} p_2^{\min(\alpha_2,\beta_2)} p_3^{\min(\alpha_3,\beta_3)} \ldots p_k^{\min(\alpha_k,\beta_k)}$	*By definition of greatest common divisor.*
$= p_1^{\max(\alpha_1,\beta_1)} p_1^{\min(\alpha_1,\beta_1)} p_2^{\max(\alpha_2,\beta_2)} p_2^{\min(\alpha_2,\beta_2)}$ $\times p_3^{\max(\alpha_3,\beta_3)} p_3^{\min(\alpha_3,\beta_3)} \ldots p_k^{\max(\alpha_k,\beta_k)} p_k^{\min(\alpha_k,\beta_k)}$ $= p_1^{\alpha_1+\beta_1} p_2^{\alpha_2+\beta_2} p_3^{\alpha_3+\beta_3} \ldots p_k^{\alpha_k+\beta_k}$ $= xy$	*We use the following equation: $\max(\alpha, \beta) + \min(\alpha, \beta) = \alpha + \beta$.*

Exercise 1F

1 **a** State the Fundamental Theorem of Arithmetic.

b Write 75 240 as a product of primes.

2 Let $n \in \mathbb{Z}^+$, $n = p_1^{a_1} p_2^{a_2} p_3^{a_3} \ldots p_k^{a_k}$.
Show that for all $1 \le i \le k$, a_i is even $\Leftrightarrow$ n is a perfect square.
Hence show that $\sqrt{20}$ is irrational.

3 Given that $n \in \mathbb{Z}^+$ and $n \times 10! = \dfrac{21!}{13!}$ show that $17 \mid n$ and $19 \mid n$.

4 Find the smallest value of a such that $7! \mid a^2$.

5 **a** Find the smallest value of $n \in \mathbb{Z}^+$ such that $2940n$ is a perfect cube.

b Determine $\sqrt[3]{2940n}$.

6 a Given that $p, q \in \mathbb{Z}^+$ such that $\gcd(p, q) = G$ and $\text{lcm}(p, q) = L$, show that $pq = GL$.

b Given that $pq = 2^4 \times 3^4 \times 5^3 \times 7^2 \times 11^3 \times 13^3$ and the least common multiple of p and q is $2^2 \times 3^3 \times 5^2 \times 7 \times 11^2 \times 13$, find the greatest common divisor of p and q.

Review exercise

EXAM-STYLE QUESTIONS

1 Consider the integers $m = 1199$ and $n = 781$, given in base 10.

a Express m and n in base 11.

b Hence show that $\gcd(781, 1199) = 11$.

2 Consider the set of numbers S of the form $n^2 + n + 17$, $n \geq 0$.

a Prove that all the elements of S are odd.

b List the first 10 elements of S.

c Show that not all the elements of S are prime.

3 Use the principle of strong mathematical induction to prove that

$2^n < \begin{pmatrix} 2n \\ n \end{pmatrix} < 4^n$ for all $n \in \mathbb{Z}^+$, $n \geq 2$.

4 a Let $a, b \in \mathbb{Z}^+$. Prove that if $a^2 \mid b^2$ then $a \mid b$.

b Is it also true that if $a^3 \mid b^3$ then $a \mid b$?

5 The sequence $\{u_n\}$, $n \in \mathbb{Z}^+$, $n \geq 2$, satisfies the recurrence relation $u_{n+1} = 7u_n - 12u_{n-1}$. Given that $u_1 = 1$ and $u_2 = 7$, use the principle of strong mathematical induction to show that $u_n = 4^n - 3^n$.

6 Let $\gcd(a, b) = d$. Show that $d \mid a + b$ and $d \mid a - b$. Hence show that $\gcd(a, a + 1) = 1$.

7 a Using the Euclidean algorithm, show that $\gcd(357, 79) = 1$.

b Find the general solution to the Diophantine equation $357x - 79y = 1$.

8 a A shop owner wants to buy the latest two video games for his shop. He has exactly \$1770 to spend. Video game A costs \$31 and game B costs \$21. What are the possible number of video games A and B that the shop owner can buy?

b He intends to sell video game A for \$45 and video game B for \$35. How many video games of each type should he buy to maximize his profits?

9 Show that if $a, b \in \mathbb{Z}^+$ and a, b are relatively prime, then $a \mid c$ if $a \mid bc$.

10 Show that if $a, b, c \in \mathbb{Z}^+$ and a, b are relatively prime, then $ab \mid c$ if $a \mid c$ and $b \mid c$.

11 Show that $\gcd(a, a + k)$ also divides k. Hence show that $\gcd(a, a + 2)$ is either 1 or 2.

12 Prove by mathematical induction that
$a^k - b^k = (a - b)(a^{k-1} + a^{k-2}b + \ldots + ab^{k-2} + b^{k-1})$ for all $k \in \mathbb{Z}^+$, $k \geq 2$.

Chapter 1 summary

Definition: A number N in base b notation is represented by $N = (d_n d_{n-1} d_{n-2} \ldots d_1 d_0)_b$ where $d_i \in \mathbb{Z}^+, 0 \le d_i < b$. The **value** of N in base b is given by:
$N = d_n \times b^n + d_{n-1} \times b^{n-1} + d_{n-2} \times b^{n-2} + \ldots + d_1 \times b^1 + d_0 \times b^0$

Definition: If $a, b \in \mathbb{Z}$, $a \neq 0$, we say that a divides b if there exists $c \in \mathbb{Z}$ such that $b = ac$. We then say that a is a **factor** of b, and b is a **multiple** of a. The notation $a|b$ denotes that a divides b.

Theorem 1:

Let $a, b, c \in \mathbb{Z}$, $a \neq 0$. Then

i $a|b$ and $a|c \Rightarrow a|(b + c)$

ii $a|b \Rightarrow a|bc$

iii $a|b$ and $b|c \Rightarrow a|c$

Theorem 2: The division theorem

Let $a \in \mathbb{Z}$ and $d \in \mathbb{Z}^+$. Then there are unique integers q and r, $0 \le r < d$, such that $a = dq + r$. We call d the divisor, q the quotient and r the remainder.

Definition: Given that $a, b \in \mathbb{Z}^+$ we say that d is the **greatest common divisor** of a and b, denoted by gcd(a,b) provided that:

i $d|a$ and $d|b$

ii if $c|a$ and $c|b$ then $c \le d$

Theorem 3:

Let $a, b \in \mathbb{Z}^+$, $a > b$, $d = \gcd(a, b)$. Then we can find $m, n \in \mathbb{Z}$ such that $d = ma + nb$.

Definition: A **linear Diophantine equation** in two variables is an equation of the form $ax + by = c$, where $a, b, c \in \mathbb{Z}$ and which has integer solutions x and y.

Theorem 4:

A linear Diophantine equation $ax + by = c$, where $a, b, c \in \mathbb{Z}$ has integer solutions in x and $y \in \mathbb{Z} \Leftrightarrow \gcd(a, b)|c$.

Corollary:

Given the linear Diophantine equation $ax + by = c$, where $a, b, c \in \mathbb{Z}$, if $\gcd(a,b) \nmid c$ there are no integer solutions to the equation.

Theorem 5:

If gcd(a, b) = d, then $\gcd\left(\frac{a}{d}, \frac{b}{d}\right) = 1$.

Theorem 6:

Given $ax + by = c$ where $\gcd(a,b) \mid c$, and x_0, y_0 is a particular solution then $\left\{x_0 + \left(\frac{b}{d}\right)k, y_0 - \left(\frac{a}{d}\right)k, k \in \mathbb{Z}\right\}$ is a complete set of solutions of the given Diophantine equation.

Definition: A positive integer p greater than 1 is said to be **prime** if the only positive factors of p are 1 and p itself. A positive integer greater than 1 which is not prime is said to be **composite**.

Theorem 7:

If n is a composite positive integer, then n has a factor less than or equal to $\sqrt{n}$.

Definition: Two integers a and b are said to be **relatively prime** or **co-prime** if gcd(a, b) = 1.

Theorem 8 (Euclid's statement):

There are infinitely many prime numbers.

Proof by Strong Mathematical Induction

The format of a proof is as follows:

- State the claim that is being proved P_n.
- Prove that the claim is true for an initial value of n, for example when $n = 1$ (though n does not necessarily have to be 1).
- Assume that the claim is true **for all** i, such that $1 \le i < k$.
- Show that if this assumption is true, the statement is also true for $n = k + 1$.
- Make a final statement to show that through the inductive process the statement is true for all values of n.

Theorem 9:

If p is a prime number and $a_1, a_2, a_3, a_4, \ldots, a_n \in \mathbb{Z}^+$ such that $p \mid (a_1 \times a_2 \times a_3 \times a_4 \times \ldots \times a_n)$, then $p \mid a_i$ for some a_i where $1 \le i \le n$.

Theorem 10: The Fundamental Theorem of Arithmetic

Every positive integer n greater than 1 can be written **uniquely** as a product of primes written in ascending order, i.e. $n \in \mathbb{Z}^+$, $n > 1 \Rightarrow n = p_1^{\alpha_1} p_2^{\alpha_2} p_3^{\alpha_3} \ldots p_i^{\alpha_i} \ldots p_m^{\alpha_m}$, where p_i is a prime number for all $1 \le i \le m$, $\alpha_i \in \mathbb{N}$ and $p_1 < p_2 < p_3 < \ldots < p_m$.

Definition: The **least common multiple** of two positive integers a and b, denoted by lcm(a, b) is the smallest positive integer that is divisible by both a and b. Given $a, b \in \mathbb{Z}^+$, lcm(a, b) = $m \Rightarrow a \mid m$ and $b \mid m$ and if there is another $n \in \mathbb{Z}^+$ such that $a \mid n$ and $b \mid n$ then $m \le n$.

2 Modular arithmetic and its applications

CHAPTER OBJECTIVES:

10.4 Modular arithmetic. The solution of linear congruences. Solution of simultaneous linear congruences (Chinese remainder theorem).

10.6 Fermat's Little Theorem ($a^p = a \pmod{p}$, where p is prime).

Before you start

You should know how to:

1 Express a number as a product of primes by successive division by prime numbers, e.g. $999 = 3^3 \times 37$

2 Convert a number from base 10 to binary using successive division by 2,

e.g. 145_{10} =

145	
72	1
36	0
18	0
9	0
4	1
2	0
1	0

$145_{10} = 10010001_2$

3 Use mathematical induction to prove that $n^3 + 2n$ is divisible by 3 for all $n \in \mathbb{Z}^+$, e.g.

$P_n: n^3 + 2n = 3A,\ A \in \mathbb{Z}^+$.

When $n = 1$, LHS $= 3 = 3 \times 1$ $\therefore P_1$ is true.

Assume P_k is true for some $k \geq 1,\ k \in \mathbb{Z}^+$.

$\Rightarrow k^3 + 2k = 3A,\ A \in \mathbb{Z}^+ \Rightarrow k^3 = 3A - 2k$

When $n = k + 1$,

LHS $= (k + 1)^3 + 2(k + 1)$

$= k^3 + 3k^2 + 3k + 1 + 2k + 2$

$= 3(A + k^2 + k + 1)$

and since $k, A \in \mathbb{Z}^+$,

then $A + k^2 + k + 1 \in \mathbb{Z}^+$.

Since we proved that P_1 was true and we showed that P_{k+1} is true whenever P_k is true, it follows by the principle of mathematical induction that P_n is true for all $n \geq 1$.

Skills check:

1 Find the prime factorization of:

a 289 **b** 8!

c 1771561 **d** 30030

2 Convert these numbers from base 10 to binary.

a 85 **b** 127

c 351 **d** 2^6

3 Use the principle of mathematical induction to prove that $n^2 - 1$ is divisible by 8 whenever n is an odd positive integer.

From Gauss to cryptography

In this chapter you will be introduced to modular congruences and modular arithmetic. The notation for congruence, ≡, first appeared in Gauss' work *Disquisitiones Arithmeticae* published in 1801, which is divided into seven sections. The first three sections of this work are about number congruences and the opening words of the first section are: "*if a number **a** divides the difference of the numbers **b** and **c**, **b** and **c** are said to be congruent with respect to **a**: but if not, incongruent. We call **a** the modulus. We shall denote in future the congruence of two numbers by the sign ≡, and adjoin the modulus in parenthesis when necessary*". Little did Gauss realize that his work would be so powerful in the age of the internet and information security.

Modular arithmetic is the basis for RSA encryption. RSA stands for Rivest, Shamir and Adelman, the names of the originators of this public-key cryptographic system. The system is said to be public-key because the algorithm for encrypting the message is publicly known, but only the person who sets up the system knows the decryption algorithm. Public-key cryptography may be compared to making a box available to the public. The person who wants to send you an encrypted message puts the message in the box and when this person sends you the closed box, only you will be able to open it with your own private key.

Sophie Germain (1776–1831) was a French mathematician who dedicated her life to the study of mathematics. In spite of opposition from her parents, she pursued her studies on her own and obtained lecture notes from the École Polytechnique. Because of the bias against female mathematicians at the time, she wrote many of her earlier works under the pseudonym Monsieur Le Blanc. She initially corresponded with Lagrange but soon her list of correspondents grew to include Legendre and Gauss. She had developed a thorough understanding of the methods presented in Gauss' *Disquisitiones Arithmeticae*. Gauss praised M. Le Blanc, and when in 1806 he found out that he had been corresponding with a woman his praise for her increased. Germain's contributions to number theory were recognized as being outstanding. In fact, she is considered to be one of the forerunners towards a proof of Fermat's Last Theorem with her studies in what are now known as Germain primes. Fermat's Last Theorem states that there are no natural number solutions to a Diophantine equation of the form $x^n + y^n = z^n$ for $n > 2$. Sophie Germain studied the equation $x^p + y^p = z^p$ where p is a prime number greater than 2 and such that $2p + 1$ is also a prime number. Her work remained the most important contribution to Fermat's Last Theorem from 1738 until 1840 when Ernst Kummer came into the picture. Sophie Germain never married and despite early opposition to her work, her father supported her financially during her lifetime.

2.1 Congruence modulo *n*

In this section we will look at natural numbers from a different perspective, by placing them in cycles. We shall first look at this informally before we delve into definitions and theorems. The tables below show you the integers 0 to 60 written in cycles of 6 and of 5.

Table A

1	2	3	4	5	6
7	8	9	10	11	12
13	14	15	16	17	18
19	20	21	22	23	24
25	26	27	28	29	30
31	32	33	34	35	36
37	38	39	40	41	42
43	44	45	46	47	48
49	50	51	52	53	54
55	56	57	58	59	60

Table B

1	2	3	4	5
6	7	8	9	10
11	12	13	14	15
16	17	18	19	20
21	22	23	24	25
26	27	28	29	30
31	32	33	34	35
36	37	38	39	40
41	42	43	44	45
46	47	48	49	50
51	52	53	54	55
56	57	58	59	60

Investigation

- What can you say about the numbers that are multiples of 6 in table A?
- What is the connection between the numbers in the second column of table B?
- Choose any two numbers from the same column and subtract the smaller number from the larger number. What do you notice?
- What happens if you add two consecutive numbers in the first column of table B?
- What if the numbers you add in the first column are not consecutive?
- Is this also true when adding numbers that are in any other column, i.e. both in the second column or third column etc…?
- Do the observations also hold for numbers in table A?
- What happens when you add numbers from different columns?
- The addition table below has been constructed to connect sums of numbers in table A. Copy and complete the table:

Addition	Number in column 1	Number in column 2	Number in column 3	Number in column 4	Number in column 5	Number in column 6
Number in column 1	*Answer is in column 2*					
Number in column 2						
Number in column 3						*Answer is in column 3*
Number in column 4						
Number in column 5			*Answer is in column 2*			
Number in column 6						

- Construct an addition table for table B.
  What conclusions can you draw from your results?

Two numbers that are in the same column of table A are said to be *congruent modulo 6*, whereas numbers that are in the same column of table B are said to be *congruent modulo 5.*

Definition

If $a \in \mathbb{Z}$ and $n \in \mathbb{Z}^+$, $n > 1$, then the **remainder** when a is divided by n is denoted by $r \equiv a(\text{mod } n)$.

All the numbers that are in column 2 of table A leave a remainder of 2 when divided by 6 and we say that they are congruent to $2(\text{mod } 6)$.

From the investigation you would have noticed that when you choose any two numbers from the same column in table A and subtract them, the answer is always divisible by 6; if you choose numbers from the same column in table B and subtract them, the answer is divisible by 5. This leads to the next definition of modular congruence.

Definition

If $a, b \in \mathbb{Z}$ and $n \in \mathbb{Z}^+$, $n > 1$, then a is **congruent** to b modulo n if n divides $a - b$. We use the following notation:
$a \equiv b(\text{mod } n) \Leftrightarrow n \mid (a - b)$

Theorem 1

Given $n \in \mathbb{Z}^+$, $a \equiv b(\text{mod } n) \Leftrightarrow a = b + kn$, where $k \in \mathbb{Z}$.

Since the statement contains a double implication we need to prove the implications in both directions.

Proof:

$\Rightarrow$:

$a \equiv b(\text{mod } n) \Rightarrow n \mid (a - b) \Rightarrow a - b = kn \Rightarrow a = b + kn$, where $k \in \mathbb{Z}$.

$\Leftarrow$:

Suppose there exists $k \in \mathbb{Z}$ such that

$a = b + kn \Rightarrow a - b = kn \Rightarrow n \mid (a - b) \Rightarrow a \equiv b(\text{mod } n)$ Q.E.D.

Let's look more closely at table A on page 42. Since each column represents $x(\text{mod } 6)$ with $x \in \{0, 1, 2, 3, 4, 5\}$, we see that $x \equiv x(\text{mod } 6)$ since $6 \mid x - x$. We say that congruence modulo 6 is **reflexive**.

Any two numbers in the same column are congruent to each other modulo 6, e.g.

$$28 \equiv 4(\text{mod } 6) \text{ and } 52 \equiv 4(\text{mod } 6)$$

$$\Rightarrow 28 \equiv 52(\text{mod } 6) \text{ and } 52 \equiv 28(\text{mod } 6)$$

We can do this for any pair of numbers in the same column; this leads to the conclusion that the relation congruence modulo 6 is **symmetric**.

Similarly, if you take any three numbers in a column, you'll see that they are all related to each other, e.g.

$$\left.\begin{aligned} 59 &\equiv 35(\text{mod } 6) \\ 35 &\equiv 11(\text{mod } 6) \end{aligned}\right\} \Rightarrow 59 \equiv 11(\text{mod } 6)$$

Again, we can do this for any three numbers in a particular column so we can deduce that congruence modulo 6 is a **transitive** relation on the positive integers. In fact we can generalize these properties as follows:

Congruence modulo n is said to be an **equivalence relation** because for all $a, n \in \mathbb{Z}^+$

- $a \equiv a(\text{mod } n)$ **(reflexive)**
- $a \equiv b(\text{mod } n) \Rightarrow b \equiv a(\text{mod } n)$ **(symmetric)**
- $a \equiv b(\text{mod } n)$ and $b \equiv c(\text{mod } n) \Rightarrow a \equiv c(\text{mod } n)$ **(transitive)**

An equivalence relation divides a set into distinct disjoint sets which form a partition of that set. For example, congruence modulo 6 divides $\mathbb{Z}^+$ into 6 distinct sets of numbers. Each set contains numbers related to each other; in this case they leave the same remainder when divided by 6. We call these sets equivalence classes.

The investigation should also have alerted you to the properties of modular congruences stated in the following theorem.

Theorem 2

If $n \in \mathbb{Z}^+$ and a, b, c and $d \in \mathbb{Z}$, $a \equiv b(\text{mod } n)$ and $c \equiv d(\text{mod } n)$ then:

i $a + c \equiv b + d(\text{mod } n)$
ii $ac \equiv bd(\text{mod } n)$

Proofs:

i Using the definition, there exist $p, q \in \mathbb{Z}$ such that

$$\left.\begin{aligned} a \equiv b(\text{mod } n) &\Rightarrow a - b = pn,\ p \in \mathbb{Z} \\ c \equiv d(\text{mod } n) &\Rightarrow c - d = qn,\ q \in \mathbb{Z} \end{aligned}\right\} \Rightarrow (a + c) - (b + d) = (p + q)n$$

Therefore $n \mid (a + c) - (b + d) \Rightarrow a + c \equiv b + d(\text{mod } n)$ Q.E.D.

ii Using the definition, there exist $p, q \in \mathbb{Z}$ such that

$$\left.\begin{aligned} a \equiv b(\text{mod } n) &\Rightarrow a = b + pn,\ p \in \mathbb{Z} \\ c \equiv d(\text{mod } n) &\Rightarrow c = d + qn,\ q \in \mathbb{Z} \end{aligned}\right\} \Rightarrow ac = (b + pn)(d + qn)$$

$\Rightarrow ac = bd + n(bq + pd + pqn)$, where $(bq + pd + pqn) = k \in \mathbb{Z}$
Therefore $n \mid (ac - bd) \Rightarrow ac \equiv bd(\text{mod } n)$. Q.E.D.

If you let $c = a$ and $d = b$ it follows using **ii** that $a^2 \equiv b^2$ (mod n).

Corollary

i $a \equiv b(\text{mod } n) \Rightarrow ka \equiv kb(\text{mod } n)$
ii $a \equiv b(\text{mod } n) \Rightarrow a^k \equiv b^k(\text{mod } n)$

These two properties are very useful and the proof using theorem 2 and mathematical induction on $k \in \mathbb{Z}$ is left as an exercise.

A direct proof of property **ii** appears later on in this chapter.

Example 1

Determine whether each of these integers is congruent to $4 \pmod 7$. **a** 80 **b** 103 **c** 326 **d** 762 **e** −32	
a $80 = 11 \times 7 + 3 \Rightarrow 80 \equiv 3 \pmod 7 \not\equiv 4 \pmod 7$	*Use the division algorithm* $n = p \times q + r$ *to find the remainder.*
b $103 = 14 \times 7 + 5 \Rightarrow 103 \equiv 5 \pmod 7 \not\equiv 4 \pmod 7$	
c $326 = 46 \times 7 + 4 \Rightarrow 326 \equiv 4 \pmod 7$	The symbol $\not\equiv$ means 'is not congruent to'.
d $762 = 108 \times 7 + 6 \Rightarrow 762 \equiv 6 \pmod 7 \not\equiv 4 \pmod 7$	
e $-32 = -5 \times 7 + 3 \Rightarrow -32 \equiv 3 \pmod 7 \not\equiv 4 \pmod 7$	

Example 2

The computer data system of a company assigns record numbers to its employees using their individual social security numbers. To avoid using very large numbers, a hashing function is used as follows: $h(N) \equiv N \pmod{163}$, where N is the social security number. **a** Find the record numbers for employees with social security numbers: **i** 178348625 **ii** 679542986 **b** What is the problem with this hashing function?	
a i $h(178348625) \equiv 178348625 \pmod{163} = 56$ **ii** $h(679542986) \equiv 679542986 \pmod{163} = 61$	*Find the remainder when dividing the social security number by 163.*
b There are only 163 record numbers available, thus there is no guarantee that every social security number will give a unique record number.	

In the next example you will see the proof for a result also known as the **cancellation theorem.**

Example 3

Given that $\gcd(n, c) = 1$, show that if $ac \equiv bc \pmod n$ then $a \equiv b \pmod n$.	
$ac \equiv bc \pmod n$ $\Rightarrow n \mid (ac - bc)$ $\Rightarrow n \mid c(a - b)$	*Use definition of linear congruence.*
$n \nmid c \Rightarrow n \mid (a - b)$ $\Rightarrow a \equiv b \pmod n$	*Since* $\gcd(n, c) = 1$.

Example 4 uses modular arithmetic to show the property of divisibility by 9. This property was introduced in chapter 1.

Example 4

Let $x_n x_{n-1} x_{n-2} \ldots x_1 x_0$ represent the number x in base 10. Show that $x \equiv x_0 + x_1 + \ldots + x_n \pmod 9$.	
We are required to show that $9 \mid x - (x_0 + x_1 + \ldots + x_n)$	
$x = x_0 \times 10^0 + x_1 \times 10^1 + \ldots + x_n \times 10^n$	*Write x as a polynomial.*
$\Rightarrow x - (x_0 + x_1 + \ldots + x_n)$ $= x_0 \times 10^0 + x_1 \times 10^1 + \ldots + x_n \times 10^n - (x_0 + x_1 + \ldots + x_n)$ $= x_0(10^0 - 1) + x_1(10^1 - 1) + \ldots + x_n(10^n - 1)$	*Use definition of modular congruence.*
Now we know that for all $k \in \mathbb{Z}^+$, $10^k - 1 = \underbrace{999\ldots.9}_{k}$	
$x - (x_0 + x_1 + \ldots + x_n) \Rightarrow$ $x = x_0(0) + x_1(9) + x_2(99) + \ldots + x_n(\underbrace{99\ldots9}_{n})$	
$\Rightarrow 9 \mid x - (x_0 + x_1 + \ldots + x_n)$ Therefore $x \equiv x_0 + x_1 + \ldots + x_n \pmod 9$	

Exercise 2A

1 Evaluate the following quantities.

a $176 \pmod 7$ **b** $-28 \pmod 5$ **c** $3501 \pmod{17}$

2 Determine whether or not each of the following integers is congruent to $5 \pmod 6$.

a 81 **b** 68 **c** −215 **d** 6785 **e** 1128

3 Consider the simultaneous equations

$$3x + y - 7z = a$$
$$2x - y + 3z = b$$
$$x + 11y - 3z = c$$

where $x, y, z, a, b, c \in \mathbb{Z}$. Show that $2a + 2b + c \equiv 0 \pmod{11}$.

4 Show that if $a \equiv b \pmod n$ and $c \equiv d \pmod n$ where $a, b, c, d \in \mathbb{Z}$ and $n \in \mathbb{Z}^+$, $n \geq 2$, then $a - c \equiv b - d \pmod n$.

5 Show that if $a \equiv b \pmod n$ where $a, b, c \in \mathbb{Z}$ and $n \in \mathbb{Z}^+$, $n \geq 2$, then $ac \equiv bc \pmod{nc}$.

6 Show that $ac \equiv bc \pmod n$ where $a, b, c \in \mathbb{Z}$ and $n \in \mathbb{Z}^+$, $n \geq 2$ does not necessarily imply that $a \equiv b \pmod n$.

7 **a** Given that $a, b \in \mathbb{N}$ and $c \in \mathbb{Z}^+$, show that if $a \equiv 1 \pmod c$, then $ab \equiv b \pmod c$.

b Using mathematical induction, show that $5^n \equiv 1 \pmod 4$.

c The positive integer N is expressed in base 5. Show that N is divisible by 4 if the sum of its digits is divisible by 4.

8 Books are identified by a ten-digit book number, $a_1a_2a_3\ldots a_{10}$ known as an ISBN. A number is a valid ISBN provided that $10a_1 + 9a_2 + 8a_3 + \ldots + 2a_9 + a_{10} = 0 \pmod{11}$.

a Is the number 0070380457 a valid ISBN?

b Find the value of x, the last digit of 071313661x, which makes this a valid ISBN number.

9 a Show that if $a \equiv b \pmod n$ and $c \equiv d \pmod n$ where $a, b, c, d \in \mathbb{Z}$ and $n \in \mathbb{Z}^+$, $n \geq 2$, then $ac = bd \pmod n$.

b Hence or otherwise show, without carrying out any long multiplication, that $35\,678 \times 25\,322 \equiv 2 \pmod 6$.

10 a Prove by induction or otherwise that for all $n \in \mathbb{N}$, $10^n - (-1)^n$ is divisible by 11.

b Let $x_n x_{n-1} x_{n-2} \ldots x_1 x_0$ represent the number x in base 10. Show that $x \equiv x_0 - x_1 + x_2 - \ldots + (-1)^n x_n \pmod{11}$.

c Hence show that 172 489 leaves a remainder of 9 when divided by 11.

11 a Define what is meant by the statement $a \equiv b \pmod n$, where $a, b, n \in \mathbb{Z}^+$.

b Hence prove that if $a \equiv b \pmod n$ then $a^3 \equiv b^3 \pmod n$.

c Determine whether the converse is always true, i.e. $a^3 \equiv b^3 \pmod n \Rightarrow a \equiv b \pmod n$.

2.2 Modular inverses and linear congruences

Suppose you are asked to find all the values of x which satisfy the linear congruence $x \equiv 2 \pmod 5$.

$x \equiv 2 \pmod 5 \Rightarrow 5 \mid x - 2 \Rightarrow x - 2 = 5k,\ k \in \mathbb{Z}$

Then $x = 5k + 2$, $k \in \mathbb{Z}$ is a general solution of this linear equation. Note that there are infinite solutions to this equation, namely $\{\ldots -13, -8, -3, 2, 7, 12 \ldots\}$.

Generalizing, we can solve the linear congruence $x \equiv a \pmod n$, where $a, n \in \mathbb{Z}^+$ as follows:
$x \equiv a \pmod n \Rightarrow x - a = kn,\ k \in \mathbb{Z}$

Therefore a general solution will be $x = kn + a$, $k \in \mathbb{Z}$.

Definition

A congruence of the form $ax \equiv b \pmod n$ where $a, b \in \mathbb{Z}$, $n \in \mathbb{Z}^+$ and x is an integer is called a **linear congruence**.

Before we move on to solving linear congruences we need to prove the following theorem.

Theorem 3

If $a, n \in \mathbb{Z}^+$, $n > 1$ where a and n are relatively prime, it follows that an inverse of $a \pmod n$, denoted by $\bar{a}$, exists such that $\bar{a}a \equiv a\bar{a} \equiv 1 \pmod n$, $\bar{a} \in \mathbb{Z}^+$, $\bar{a} < n$. Furthermore, this inverse is unique modulo n.

We described how to solve the special case of linear congruence just before the definition.

Proof:

The proof is done in two parts. First we must prove that the inverse exists and then we must show that it is unique.

i Existence

In chapter 1 we showed that if $\gcd(a, n) = 1$ we can find integers x and y such that $xa + yn = 1$.
Since a and n are relatively prime it follows that:

$\gcd(a, n) = 1 \Rightarrow xa + yn = 1 \Rightarrow xa + yn \equiv 1 \pmod n$.

But $yn \pmod n \equiv 0 \pmod n \Rightarrow xa + yn \equiv xa \pmod n$.

Combining these results, it follows that $xa \equiv 1 \pmod n$ and so x is an inverse of $a \pmod n$.

ii Uniqueness

Let's now assume that this inverse is not unique, i.e. there is some $t \in \mathbb{Z}^+$ such that $ta \equiv xa \pmod n$ and $t \not\equiv x \pmod n$.

Then $n \mid ta - xa \Rightarrow n \mid a(t - x)$.

Since we know that a and n are co-prime, it follows that $n \mid (t - x)$.

$\Rightarrow t - x = kn, k \in \mathbb{Z}^+ \Rightarrow t \equiv x \pmod n$, which is a contradiction.
Therefore the multiplicative inverse is unique (mod n). Q.E.D.

Example 5 will show you how to find the multiplicative inverse of a given number (mod n).

In question 4 of Exercise 2A you proved that if $a \equiv b \pmod n$ and $c \equiv d \pmod n$, where $a, b, c, d \in \mathbb{Z}$ and $n \in \mathbb{Z}^+$, $n \geq 2$, then $a - c \equiv b - d \pmod n$.
We shall now use this result to prove the next theorem.

Theorem 4

For all $n, k \in \mathbb{Z}^+$, $n > k$, $na - k \pmod n \equiv -k \pmod n$.

Proof:

By definition we know that $na \equiv 0 \pmod n$ since $n \mid na - 0$.
Since $k < n$ and $n \mid (k - k)$ we also have that $k \equiv k \pmod n$.
Combining these two results we obtain
$na - k \equiv 0 - k \pmod n \Rightarrow na - k \equiv -k \pmod n$. Q.E.D.

For example, $69 \equiv 6 \pmod 7 \equiv -1 \pmod 7$ since $6 = 7 - 1$.

This result is useful when working out congruences as demonstrated in the next examples.

Example 5

In each part, determine whether a multiplicative inverse exists, and if it does find it.

a $3 \pmod 7$ **b** $34 \pmod{51}$ **c** $7 \pmod{24}$ **d** $8 \pmod{51}$

a **Either**	
Since $\gcd(3, 7) = 1$ an inverse exists.	*Use the Euclidean algorithm to find gcd (7, 3).*
$1 = 7 - 2 \times 3$	*Find x, y such that* $1 = 7x + 3y$.
Therefore the multiplicative inverse of $3 \pmod 7$ is -2 or 5.	
OR	
$3 \equiv 3 \pmod 7$ $5 \times 3 \equiv 15 \pmod 7 \equiv 1 \pmod 7$ Therefore the multiplicative inverse of $3 \pmod 7$ is 5 or -2.	*When the numbers are small it is sometimes easier to find a multiplier that leaves a remainder of 1.*
b $51 = 34 \times 1 + 17$ $34 = 17 \times 2 + 0$ $\gcd(51, 34) \neq 1$, therefore $34 \pmod{51}$ does not have an inverse.	*Use the Euclidean algorithm to find gcd (51, 34).*
c $24 = 7 \times 3 + 1$ $7 = 3 \times 2 + 1$ Since $\gcd(24, 7) = 1$ an inverse exists.	*Use the Euclidean algorithm to find gcd (7, 24).*
Either $1 = 7 - 3 \times 2$ $= 7 - 2(24 - 7 \times 3)$ $= 7 \times 7 - 2 \times 24$	*Work backwards to find x, y such that* $1 = 7x + 24y$.
Therefore the multiplicative inverse of $7 \pmod{24}$ is 7.	*7 is a self inverse in modulo 24.*
OR $7 \equiv 7 \pmod{24}$ $7 \times 7 \equiv 49 \pmod{24} \equiv 1 \pmod{24}$ Therefore multiplicative inverse of 7 is 7.	
d $51 = 8 \times 6 + 3$ $8 = 3 \times 2 + 2$ $3 = 2 \times 1 + 1$	*Use the Euclidean algorithm to find gcd (8, 51).*
Since $\gcd(8, 51) = 1$ an inverse exists. $1 = 3 - 2 \times 1$ $= 3 - (8 - 3 \times 2)$ $= 3 \times 3 - 1 \times 8$	*Work backwards to find x, y such that* $1 = 51x + 8y$.
$= 3(51 - 6 \times 8) - 1 \times 8$ $= 3 \times 51 - 19 \times 8$ Therefore the multiplicative inverse of $8 \pmod{51}$ is -19 or 32.	*Note that in this example it is easier to work backwards than to find a multiple of 8 that leaves a remainder of 1 when divided by 51.*

The next example demonstrates how to use the multiplicative inverse to solve linear congruences.

Example 6

Find the solutions to the linear congruences:

a $3x \equiv 5(\text{mod } 7)$ **b** $9x \equiv 3(\text{mod } 5)$ **c** $287x \equiv 3\text{mod}(319)$

a From Example 5 part **a**, we know that the multiplicative inverse of $3(\text{mod } 7)$ is -2.	*Show that 7 and 3 are co-prime and find a, b such that $1 = 7a + 3b$.*
Hence $-2 \times 3x \equiv -2 \times 5(\text{mod } 7)$ $\Rightarrow -6x \equiv -10(\text{mod } 7)$	*Multiply both sides by the inverse of 3 (mod 7).*
$\Rightarrow -6(\text{mod } 7) \times x(\text{mod } 7) \equiv -10(\text{mod } 7)$ $\Rightarrow x \equiv -3(\text{mod } 7)$	*$-6(\text{mod } 7) = 1$.*
$\Rightarrow x \equiv 4(\text{mod } 7)$	*Using Theorem 2.*
Therefore $x = 7k + 4,\ k \in \mathbb{Z}$. (5 is also a multiplicative inverse of 3(mod 7). It is left to you to confirm that the same result is obtained when using 5 instead of –2.)	*Write the general solution.*
b $9x \equiv 3(\text{mod } 5)$ $9 \equiv -1(\text{mod } 5)$ $\Rightarrow -9 \equiv 1(\text{mod } 5)$ Therefore the inverse of $9(\text{mod } 5)$ is -1.	
$-9x \equiv -3(\text{mod } 5)$ $\Rightarrow x \equiv 2(\text{mod } 5)$	*Multiply both sides of the equation by the inverse.*
Therefore $x = 5k + 2,\ k \in \mathbb{Z}$.	*Write the general solution.*
c $319 = 287 \times 1 + 32$ $287 = 32 \times 8 + 31$ $32 = 31 \times 1 + 1$ $\gcd(319, 287) = 1$ $1 = 32 - 1 \times 31$ $= 32 - 1(287 - 8 \times 32)$ $= 9 \times 32 - 1 \times 287$ $= 9(319 - 1 \times 287) - 1 \times 287$ $= 9 \times 319 - 10 \times 287$	*Use the Euclidean algorithm to show that 319 and 287 are co-prime and find a, b such that $1 = 287a + 319b$.*
Therefore the inverse of $287(\text{mod } 319)$ is -10.	*Identify the inverse of 287 (mod 319).*
$287x \equiv 3\text{ mod}(319)$ $\Rightarrow -2870x \equiv -30(\text{mod } 319)$ $\Rightarrow x \equiv 289(\text{mod } 319)$	*Multiply both sides of the congruence by the inverse and solve for x.*
Therefore $x = 319k + 289,\ k \in \mathbb{Z}$.	

Example 7

Without carrying out any addition or multiplication, find the remainder when: **a** 1235 + 4684 is divided by 12 **b** 252 × 168 is divided by 11 **c** 2345 × 7013 is divided by 23	
a $1235 \equiv -1 \pmod{12}$ $4684 \equiv 4 \pmod{12}$	*$1235 = 12 \times 102 + 11$* *$4684 = 12 \times 390 + 4$*
$1235 + 4684 \equiv -1 + 4 \pmod{12}$ $\equiv 3 \pmod{12}$ Therefore the remainder is 3.	
b $252 \equiv -1 \pmod{11}$ $168 \equiv 3 \pmod{11}$	*When 252 is divided by 11 the remainder is 10.* *The remainder when 168 is divided by 11 is 3.*
$252 \times 168 \equiv -3 \pmod{11} \equiv 8 \pmod{11}$ Therefore the remainder is 8.	*Remainder $r < 0$, so use theorem 4 to change to a positive integer.*
c $2345 \equiv -1 \pmod{23}$ $7013 \equiv -2 \pmod{23}$	
$2345 \times 7013 \equiv -1 \times -2 \pmod{23}$ $\equiv 2 \pmod{23}$ Therefore the remainder is 2.	*2345 leaves a remainder of 22 and 7013 leaves a remainder of 21 when divided by 23.*

Exercise 2B

1 For each of the following, determine whether or not a multiplicative inverse exists, and if so then find it.

a $5 \pmod{21}$ **b** $35 \pmod{63}$

c $108 \pmod{153}$ **d** $17 \pmod{50}$

2 Without carrying out any addition or multiplication, find the remainder when:

a 1632 + 2467 is divided by 11

b 3715 × 2369 is divided by 21

c 784 × (566 + 723) is divided by 15

3 Find the solutions to these linear congruences:

a $13x \equiv 4 \pmod{28}$

b $156x \equiv 11 \pmod{71}$

c $108x \equiv 2 \pmod{133}$

4 a If k, $0 \leq k < 11$, is a solution of the congruence $4x \equiv 5 \pmod{11}$, find the value of k.

b Show that all solutions of $4x \equiv 5 \pmod{11}$ are congruent to k.

5 Let the greatest common divisor of 1192 and 1108 be d.

a Using the Euclidean algorithm, find d.

b Hence find integers a and b such that $1192a + 1108b = d$.

c Using part **b**, solve $277x \equiv 2 \pmod{298}$, where $x < 298$, $x \in \mathbb{Z}^+$.

2.3 The Pigeonhole Principle

It is often the case in mathematics that a simple fact can be applied in an elegant way to solve a problem that may initially seem difficult to solve. The Pigeonhole Principle is a very simple rule that is extremely useful when solving some problems involving numbers.

The Pigeonhole Principle: If m pigeons occupy n pigeonholes and $m > n$, then at least one pigeonhole must be occupied by more than one pigeon.

? Gustav Lejeune Dirichlet (1805 – 1859) is very well-known for his fundamental work on functions. However, Dirichlet was also very active in Number Theory and he was the first mathematician to formulate the Pigeonhole Principle which is sometimes referred to as the Dirichlet box principle. It was also Dirichlet who proved Fermat's Last Theorem for the special case when $n = 5$, i.e. that there are no integer solutions to the equation $x^5 + y^5 = z^5$.

As simple and obvious as the Pigeonhole Principle might seem, it can be applied in various situations as the following examples will show.

Example 8

What is the minimum number of students in a Mathematics HL class that will guarantee that at least two students

a obtain the same grade in Mathematics

b have a birthday in the same month?

a Since there are 7 possible grades, it follows using the Pigeonhole Principle that 8 is the smallest number of students that will guarantee at least two grades that are the same.

Consider the possible grades available.
Apply the Pigeonhole Principle.

Note that although 8 students will guarantee that at least two students achieved the same grade, not all grades need to be represented in the results. This means that the pigeonhole principle does not imply that all pigeonholes have to be occupied.

b There are 12 months in a year so the smallest number of students that will guarantee that at least 2 students have their birthday in the same month is 13.

Consider the number of months possible.
Apply the Pigeonhole Principle.
The number of students has to be greater than the number of months.

Example 9

Tom sorts out his building blocks according to size. In a box he has building blocks that have the same shape and size but different colours. In one of the boxes he has 50 red blocks, 35 blue, 25 green and 40 white. **a** What is the minimum number of blocks he can pick from this box (while blindfolded) to guarantee that he has picked an identical pair of blocks? **b** How many blocks must he pick to ensure that he has a pair of identical red blocks?	
a There are 4 different colours in the box. He must therefore pick 5 blocks to guarantee that 2 are of the same colour.	*First list the possible blocks he can choose from. Apply the Pigeonhole Principle since the number of colours is n and the number of blocks is m, and the condition is that m > n he must pick at least 5 blocks.*
b There are 150 blocks, 50 of which are red. Tom will have to pick 102 blocks to ensure that at least 2 blocks are red.	*Sort out blocks into red and other colours. Worse case scenario, he could pick all the non-red blocks first. So he must pick another 2 to guarantee the result.*

Example 10

How many students would need to attend the next IB student conference to ensure that there are at least two students with the same initials?	
The number of different ways of combining first and second initials is 26^2.	*There are 26 letters in the alphabet. Find the number of different ways of combining first and second initials.*
So $26^2 + 1 = 677$ students must attend the conference to guarantee that at least two students have the same initials.	*Use the Pigeonhole Principle where n is the number of possible different initials and m is the number of students.*

The following examples illustrate a more elegant use of the Pigeonhole Principle used in Number Theory.

Example 11

Let A be a set containing 25 positive integers. Show that there are two elements in A that will leave the same remainder when divided by 24.	
For all $m \in A$, $m = 24q + r$, $0 \leq r < 24$. Since there are 25 positive integers and each can be written in this form, by applying the Pigeonhole Principle there must be at least two integers that leave the same remainder when divided by 24.	*Use the division algorithm.*

Example 12

Show that there is a positive integer n, such that 11 divides $2^n - 1$.	
Let $A = \{2-1, 2^2-1, 2^3-1, \ldots, 2^{12}-1\}$.	*Write out 12 integers in the form $2^n - 1$.*
Consider $2^i - 1 \equiv a \pmod{11}$ for $0 \le i \le 12$.	*Use modular arithmetic to denote the remainders when each of these numbers is divided by 11.*
Then by the Pigeonhole Principle there must be some j, $0 \le j < i \le 12$, such that $2^i - 1 \equiv 2^j - 1 \pmod{11}$	*There are only 11 possible remainders when dividing by 11.*
$\Rightarrow 2^i \equiv 2^j \pmod{11}$	
$\Rightarrow 2^i - 2^j = 11k,\ k \in \mathbb{Z}^+$	*Since $0 \le j < i \le 12$, 2^j and $2^{i-j} - 1$ are positive integers.*
$\Rightarrow 2^j(2^{i-j} - 1) = 11k$	
$\Rightarrow 11 \mid 2^j(2^{i-j} - 1)$	*Since 11 is an odd number $\gcd(2^j, 11) = 1$.*
$\Rightarrow 11 \mid (2^{i-j} - 1)$	

Example 13

Show that if 16 integers are chosen from the set $A = \{n \mid n \in \mathbb{Z}^+, n \le 30\}$, then at least one of the chosen integers must divide another chosen integer.	
Each chosen integer can be written as $2^i k$ where k is an odd number, $k \le 29$.	*This is true using the Fundamental Theorem of Arithmetic, e.g.* $A = \{2^0 \times 1, 2^1 \times 1, 2^0 \times 3, 2^2 \times 1, \ldots, 2^0 \times 29, 2^1 \times 15\}$
The number of odd numbers k is therefore 15, and since 16 numbers are chosen there must be the same value of k for two numbers $a = 2^i k$, $b = 2^j k$, where $i > j \Rightarrow b \mid a$ and $i < j \Rightarrow a \mid b$.	*The largest number that can be chosen is 30 so the largest odd number k is 29.* *Apply the Pigeonhole Principle for k.*

The results obtained in the previous two examples can be generalized. These are left for you to prove in questions 5 and 6 of Exercise 2C.

The Pigeonhole Principle can be generalized as follows:

The Generalized Pigeonhole Principle
If n pigeonholes are occupied by $kn + 1$ or more pigeons, where $k \in \mathbb{Z}^+$, then at least one pigeonhole must be occupied by $k + 1$ or more pigeons.

This does not mean that each pigeonhole has to be occupied.

Exercise 2C

1 a There are 29 students in a class. At least how many students have first names that begin with the same letter?

b A drawer contains 10 red pens and 15 blue pens. How many pens should be drawn (while you are blindfolded) to ensure that 3 blue pens are included?

c Gino suggests that the items in a warehouse should be given an alphanumeric code consisting of a letter of the alphabet followed by two digits. His manager tells him that this will not work because some items are bound to have the same code. At least how many different items are in the warehouse?

2 Show that if any 6 integers are chosen from the set $S = \{1, 2, 3, \ldots, 10\}$, at least two of them must add up to 11.

3 a 26 numbers are chosen from the integers between 1 and 50 inclusive. Show that at least two of the integers are consecutive.

b Hence or otherwise prove that if we choose $n + 1$ integers from the set $A = \{k \mid k \in \mathbb{Z}^+, k \leq 2n\}$, then there will be at least one consecutive pair.

4 a Six positive integers are chosen at random. Show that at least two of them will leave the same remainder when divided by 5.

b Hence or otherwise prove that if any $n + 1$ positive integers are chosen at random, at least two of them will leave exactly the same remainder when divided by n.

5 Prove that if $n + 1$ integers are chosen from the set $A = \{k \mid k \in \mathbb{Z}^+, k \leq 2n\}$, at least one of the chosen integers must divide another chosen integer.

6 Show that for every positive odd number m there exists $n \in \mathbb{Z}^+$ such that m divides $2^n - 1$.

2.4 The Chinese Remainder Theorem or systems of linear congruences

孫子算經卷上
唐朝議大夫行太史令上輕車都尉臣李淳風等奉 勑注釋
度之所起起於忽欲知其忽蠶吐絲爲忽十忽
爲一絲十絲爲一毫十毫爲一氂十氂爲一分
十分爲一寸十寸爲一尺十尺爲一丈十丈爲
一引五十尺爲一端四十尺爲一疋六尺爲一
步二百四十步爲一畝三百步爲一里
稱之所起起於黍十黍爲一絫十絫爲一銖二
十四銖爲一兩十六兩爲一斤三十斤爲一鈞

In the third chapter of a Chinese book *The Mathematical Classic of Sun Zi* attributed to Sun Zi Suanjing between the third and fifth century AD, problem 26 states the following: "*Now there are an unknown number of things. If we count by threes, there is a remainder 2; if we count by fives, there is a remainder 3; if we count by sevens, there is a remainder 2. Find the number of things.*"

If we let x represent the number we are looking for, then the three conditions above can be written as a triplet of linear congruences as follows:

$$x \equiv 2 \pmod 3$$
$$x \equiv 3 \pmod 5$$
$$x \equiv 2 \pmod 7$$

Of course, we could try to find the solution by going through the positive integers that leave a remainder of 2 when divided by 7 and checking their remainders when divided by 5 and by 3. We would very quickly find out that 23 satisfies all three conditions.

If we are given a pair of simultaneous linear congruences we can solve by using substitution. The next example illustrates this.

Example 14

Solve the following pair of linear congruences:
$x \equiv 2 \pmod 7$
$3x \equiv 2 \pmod 5$

$x \equiv 2 \pmod 7 \Rightarrow x = 7t + 2$	*Using the definition of modulo 2.*
$\Rightarrow 3x = 3(7t + 2) \equiv 2 \pmod 5$	*Substitute for x in second equation.*
$\Rightarrow 21t + 6 \equiv 2 \pmod 5$	
$\Rightarrow 21t \equiv -4 \pmod 5 \equiv 1 \pmod 5$	*Subtract 6.*
$\Rightarrow 20t + t \equiv 1 \pmod 5$	
$\Rightarrow t \equiv 1 \pmod 5$	*$20t \equiv 0 \pmod 5$*
$\Rightarrow t = 5k + 1$	*Simplify.*
But	
$x = 7t + 2 = 7(5k + 1) + 2$	
$\Rightarrow x = 35k + 9, k \in \mathbb{Z}^+$	*Substitute for t and find the general solution.*

To find a solution for three or more simultaneous linear congruences a different method is required and there is an algorithm that works, provided that the divisors are pairwise co-prime. Before we prove the general algorithm let's go through the example of Sun Zi's puzzle and try to solve it step by step.

We notice that the divisors 3, 5 and 7 are pairwise co-prime, i.e. $\gcd(3, 5) = 1$, $\gcd(3, 7) = 1$ and $\gcd(5, 7) = 1$

If we find numbers A, B and C such that

- A leaves a remainder of 2 when divided by 3 but divides both B and C
- B leaves a remainder of 3 when divided by 5 but divides both A and C
- C leaves a remainder of 2 when divided by 7 but divides both A and B

then the number $x_0 = A + B + C$ will satisfy all three conditions.

We will now try to find three such numbers to help us solve Sun Zi's problem and this is where our knowledge of modular arithmetic comes in handy.

If we take the product $5 \times 7 = 35$, we know that $\gcd(35, 3) = 1$. This means that 35 has an inverse modulo 3, and we can find it.

The same applies if we take the product $3 \times 7 = 21$ where $\gcd(21, 5) = 1$ and we can find an inverse of 21 modulo 5.

And if we take the product $3 \times 5 = 15$, $\gcd(15, 7) = 1$, then 15 has an inverse modulo 7.

It is left as an exercise for you to find the inverses and check that the following are true:

$2 \times 35 \equiv 1 \pmod 3 \Rightarrow 140 \equiv 2 \pmod 3 \Rightarrow A = 140$

$1 \times 21 \equiv 1 \pmod 5 \Rightarrow 63 \equiv 3 \pmod 5 \Rightarrow B = 63$

$1 \times 15 \equiv 1 \pmod 7 \Rightarrow 30 \equiv 2 \pmod 7 \Rightarrow C = 30$

One solution for x would be $x_0 = 140 + 63 + 30 = 233$. However, it is not the smallest solution, which, we know, is 23. In fact, any multiple of 105, (the product of 3, 5 and 7) added to or subtracted from 233 will also satisfy all three congruences.

In order to find a general solution we first need to find the smallest positive integer that is a solution, and to do this we need to find the remainder when 233 is divided by 105:

$233 \equiv 23 \pmod{105}$

Therefore the general solution is $x = 23 + 105k,\ k \in \mathbb{Z}^+$.

If you were able to follow the working in this example you should be able to understand the proof of the Chinese Remainder Theorem which holds for any number of linear congruences.

Theorem 5: The Chinese Remainder Theorem

Given the system of linear congruences:

$x \equiv a_1 \pmod{m_1}$

$x \equiv a_2 \pmod{m_2}$

$x \equiv a_3 \pmod{m_3}$

$\vdots$

$x \equiv a_n \pmod{m_n}$

If $m_1, m_2, m_3, \ldots, m_n$ are pairwise relatively prime positive integers, there exists a unique solution modulo M where $M = m_1 \times m_2 \times m_3 \times \ldots \times m_n$. In other words, there is a particular solution $0 \le x_0 < M$ and a general solution given by $x \equiv x_0 + kM$.

Proof:

Let $M = m_1 \times m_2 \times \ldots \times m_n$ and let $M_k = \dfrac{M}{m_k}$.

By this definition it follows that $\gcd(M_k, m_k) = 1$, which means that M_k has an inverse p_k such that:
$M_k p_k \equiv 1 \pmod{m_k} \Rightarrow a_k M_k p_k \equiv a_k \pmod{m_k}$

Then one particular solution would be
$x_p = a_1 M_1 p_1 + a_2 M_2 p_2 + \ldots + a_k M_k p_k + \ldots + a_n M_n p_n$

You can verify that $x_p \equiv a_i (\text{mod } m_i)$ because

$$M_k = \frac{M}{m_k} \Rightarrow M_k \equiv 0(\text{mod}\, m_i), \text{ for } i \neq k \text{ and } M_k \equiv a_k(\text{mod}\, m_k).$$

We then need to find the smallest solution by taking $x_0 \equiv x_p (\text{mod } M)$ and the general solution is given by $x = x_0 + kM$. Q.E.D.

The next example illustrates how to use this algorithm for a system of three simultaneous linear congruences.

Example 15

Solve the following system of linear congruences: $x \equiv 2(\text{mod } 5)$ $x \equiv 3(\text{mod } 7)$ $x \equiv 4(\text{mod } 11)$	
$M = 385$, $M_1 = 77$, $M_2 = 55$ and $M_3 = 35$ $77 \equiv 2(\text{mod } 5)$	*Find M, M_1, M_2, M_3.* *Since the condition is already met we may stop here.*
$55 \equiv 6(\text{mod } 7)$ $6 \times 55 \equiv 1(\text{mod } 7)$ $\Rightarrow 18 \times 55 = 3(\text{mod } 7)$	*6 is the inverse of M_2(mod 7).* *Multiply by 3.*
$35 \equiv 2(\text{mod } 11)$ $6 \times 35 \equiv 1(\text{mod } 11)$ $\Rightarrow 24 \times 35 \equiv 4(\text{mod } 11)$	*6 is the inverse of M_3(mod 11).* *Multiply by 4.*
$77 + 18 \times 55 + 24 \times 35 = 1907$	
Therefore a solution would be 1907 and $x_0 = 367$. A general solution would therefore be: $x = 367 + 385k$, $k \in \mathbb{Z}^+$.	*Find the primary solution, x_0.* *Write the general solution.*

In the next example we will use the algorithm to solve the pair of linear congruences in Example 14.

Example 16

Using the algorithm of the Chinese Remainder Theorem, solve the following pair of linear congruences: $x \equiv 2 \pmod 7$ $3x \equiv 2 \pmod 5$	
$3x \equiv 2 \pmod 5$ $\Rightarrow 6x \equiv 4 \pmod 5$ $\Rightarrow x \equiv 4 \pmod 5$	*Multiply the congruence by the inverse of 3, the coefficient of x, and simplify.*
The two congruences become $x \equiv 2 \pmod 7$ $x \equiv 4 \pmod 5$	
$\gcd(5, 7) = 1$	$M = 35$
$5 \equiv 5 \pmod 7$ $\Rightarrow 15 \equiv 1 \pmod 7$ $\Rightarrow 30 \equiv 2 \pmod 7$	*3 is the inverse of 5 (mod 7).* *Multiply by 2 to obtain 30 which is the first multiple of 5 that satisfies this congruence.*
$7 \equiv 2 \pmod 5 \Rightarrow 21 \equiv 1 \pmod 5$ $\Rightarrow 84 \equiv 4 \pmod 5$	*3 is the inverse of 2 (mod 5).* *Multiply by 4 to obtain 84 which is the first multiple of 7 that satisfies this congruence.*
$30 + 84 = 114 \Rightarrow x_0 = 114 \equiv 9 \pmod{35}$	*Find x_0.*
General solution is $x = 9 + 35k,\ k \in \mathbb{Z}^+$.	*Write the general solution.*

Investigation

The Indian mathematician Brahmagupta (7th Century AD) posed the following:

"*When eggs in a basket are removed two, three, four, five or six at a time, there remain respectively one, two, three, four or five eggs. When they are taken out seven at a time none are left over. Find the smallest number of eggs that could have been contained in the basket.*"

- Write this information as a set of 6 linear congruences.
- What do you notice about the 6 linear congruences?
- Which congruences would you eliminate in order to find a solution?
- What is the solution for the remaining set of congruences?
- Does this solution also satisfy the congruences which were eliminated?

In this investigation you will have noticed that a solution could be found, however this is not always the case. Consider the following set of linear congruences:

$x \equiv 1 \pmod 2$

$x \equiv 1 \pmod 3$

$x \equiv 2 \pmod 6$

In this set of equations the divisors are not mutually co-prime and there is no solution. The equations are reduced to 3 linear Diophantine equations as follows:

$x \equiv 1 \pmod 2 \Rightarrow x - 1 = 2y \Rightarrow x - 2y = 1$

$x \equiv 1 \pmod 3 \Rightarrow x - 1 = 3y \Rightarrow x - 3y = 1$

$x \equiv 2 \pmod 6 \Rightarrow x - 2 = 6y \Rightarrow x - 6y = 2$

There are three equations and two unknowns. If you solve the equations pair-wise, the solutions will not satisfy the third equation.

The next example shows a possible application of the Chinese Remainder Theorem.

Example 17

Carlos originally had a collection of 100 toy soldiers, but he lost some of them. One day he arranged his soldiers in rows of 3 and found that the last row had only 2 soldiers. He then arranged them in rows of 4, but again the last row had only 2 soldiers. He decided to arrange them in rows of 5 and this time the last row had only 3 soldiers. How many of the original 100 toy soldiers had he lost?

$x \equiv 2 \pmod 3$ $x \equiv 2 \pmod 4$ $x \equiv 3 \pmod 5$	*Write the information given as a set of linear congruences.*
$M = 60$, $M_1 = 20$, $M_2 = 15$ and $M_3 = 12$	*Find M, M_1, M_2, M_3.*
$20 \equiv 2 \pmod 3$	
$15 \equiv 3 \pmod 4$ $\Rightarrow 30 \equiv 2 \pmod 4$	*Sometimes it is easier to find which multiple will give the remainder required thus avoiding working with large numbers.*
$12 \equiv 2 \pmod 5$ $\Rightarrow 48 \equiv 3 \pmod 5$	*Multiplying by four will give the required congruence.*
$20 + 30 + 48 = 98 \equiv 38 \pmod{60} = 278$ $= 38 \pmod{60}$ $\Rightarrow x_0 = 38$ giving a general solution $x = 38 + 60n$.	*Find the primary solution x_0.* *Write the general solution.*
But we want $0 < x \le 100$, so $x = 38$ or 98.	*Find x to satisfy the conditions given.*
Therefore, either 62 or 2 of the original toy soldiers had been lost.	

In Example 17, short cuts were taken by finding a multiple which gives the required congruence. It is left up to you to show that the answer obtained would be the same if the algorithm were followed using inverses.

Exercise 2D

1 Solve the pair of linear congruences.
$x \equiv 2 \pmod{11}$
$x \equiv 2 \pmod{13}$

2 Find the smallest positive integer that leaves a remainder of 3 when divided by 7, and a remainder of 1 when divided by 12.

3 Solve the system of linear congruences.
$x \equiv 1 \pmod{3}$
$2x \equiv 3 \pmod{5}$

4 Solve the systems of linear congruences.

a $x \equiv 3 \pmod{4}$
$x \equiv 1 \pmod{5}$
$x \equiv 2 \pmod{7}$

b $x \equiv 1 \pmod{3}$
$x \equiv 2 \pmod{4}$
$x \equiv 3 \pmod{5}$

c $x \equiv 1 \pmod{2}$
$x \equiv 3 \pmod{7}$
$x \equiv 1 \pmod{11}$

5 Marielle buys some tulip bulbs to plant in her garden. She is not sure how many bulbs she has in total, but while deciding how to plant them she notices the following: if she plants them in groups of 8 she has only 6 bulbs in the last group, and if she plants them in groups of 9 she has 8 bulbs left over. She knows that she has less than 100 bulbs. How many more bulbs would she need in order to plant all the bulbs evenly in groups of 10?

6 On his birthday party invitation to his school friends, Paul included the following puzzle:
If my age is divided by 2 or by 5 the remainder is 1, and if divided by 4 the remainder is 3. How old is Paul?

7 Karina and Yasmin are helping at the school fund-raising evening. They have a large box full of cookies that they want to pack into bags. They were told that the box contained between 300 and 350 cookies. They notice that if they pack them in bags of 3, 5 or 7 they are left with one extra cookie, but if they pack them in bags of 4 then no cookies are left over. Assuming that they did not eat any of the cookies, how many cookies did they pack?

2.5 Using cycles for powers modulo *n* and Fermat's Little Theorem

We have already shown, on page 45, that if $a \equiv b \pmod{n}$ where $a, b \in \mathbb{Z}$ and $n \in \mathbb{Z}^+$, then $a^2 \equiv b^2 \pmod{n}$. And in Exercise 2A question 11 under the same conditions you showed that $a^3 \equiv b^3 \pmod{n}$. In the next theorem we shall generalize this result.

On page 45 in the text box next to Theorem 2

Theorem 6

Given that $a \equiv b \pmod{n}$ where $a, b \in \mathbb{Z}$ and $n \in \mathbb{Z}^+$, then $a^m \equiv b^m \pmod{n}$ for all $m \in \mathbb{Z}^+$

Proof:

$a \equiv b \pmod{n} \Rightarrow a - b = kn \Rightarrow a = b + kn$

Therefore $a^m = (b + kn)^m$.

Using the binomial theorem:

$$a^m = \binom{m}{0} b^m + \binom{m}{1} b^{m-1} kn + \ldots + \binom{m}{r} b^{m-r} (kn)^r + \ldots + \binom{m}{m} (kn)^m$$

$$= b^m + \left(\binom{m}{1} b^{m-1} k + \ldots + \binom{m}{r} b^{m-r} k^r n^{r-1} + \ldots + \binom{m}{m} (k^m n)^{m-1} \right) n$$

$$\Rightarrow a^m - b^m = \left(\binom{m}{1} b^{m-1} k + \ldots + \binom{m}{r} b^{m-r} k^r n^{r-1} + \ldots + \binom{m}{m} (k^m n)^{m-1} \right) n$$

An alternative way to prove this theorem is using Theorem 2 and mathematical induction, but this is left as an exercise for you.

Therefore $n \mid a^m - b^m \Rightarrow a^m \equiv b^m \pmod{n}$. Q.E.D.

This result is very important when working with modular congruences using large powers. If you were asked to find the remainder when 25^{453} is divided by 12 you will realize that a calculator is useless because the number is too big and causes an overflow. This is where modular arithmetic comes in very handy.

$25 \equiv 1 \pmod{12} \Rightarrow 25^{453} \equiv 1^{453} \pmod{12} \equiv 1 \pmod{12}$. So the remainder when 25^{453} is divided by 12 is 1.

Similarly if we want to find the remainder when 98^{101} is divided by 11 we can argue that $98 \equiv -1 \pmod{11} \Rightarrow 98^{101} \equiv -1^{101} \pmod{11} \equiv -1 \pmod{11}$. Since by the division algorithm the remainder is always greater or equal to zero, the remainder when 98^{101} is divided by 11 is 10.

Example 18

Find the remainder when **a** 2^{51} is divided by 7 **b** 2^{21} is divided by 41.	
a $2^3 \equiv 1 \pmod 7$ $2^{51} \equiv (2^3)^{17} \equiv 1^{17} \pmod 7$ Therefore the remainder when 2^{51} is divided by 7 is 1.	*Find the smallest power of 2 which leaves a remainder when divided by 7.* *Rewrite 51 in terms of 2^3.*
b $2^{21} = 2 \times 2^{20}$ $2^5 \equiv 32 \equiv -9 \pmod{41}$ $2^{20} = (2^5)^4 \equiv (-9)^4 \pmod{41}$ $\equiv (81)^2 \pmod{41}$ Since $81 \equiv -1 \pmod{41}$, it follows that $2^{20} \equiv (81)^2 \pmod{41} \equiv (-1)^2 \pmod{41}$ $\equiv 1 \pmod{41}$. $2^{21} \equiv 2 \pmod{41} \times 2^{20} \pmod{41} = 2$ Therefore the remainder when 2^{21} is divided by 41 is 2.	*Find the smallest power of 2 that gives a number close to 41.*

A nice feature of modular arithmetic is that it allows us to compute remainders of very large numbers when written in exponent form. We have already shown how this can be done in some examples in this chapter, but what if the powers are not so obviously factorized? This is where powers of 2 come in handy.

Let's try to compute the remainder when 7^{256} is divided by 13.

$256 = 2^8$

$7^2 = 49 \equiv -3 \pmod{13}$

$7^4 = 7^{2^2} \equiv 9 \pmod{13}$

$7^8 = 7^{2^3} \equiv 81 \pmod{13} \equiv 3 \pmod{13}$

$7^{16} = 7^{2^4} \equiv 9 \pmod{13}$

$7^{32} = 7^{2^5} \equiv 3 \pmod{13}$

$7^{64} = 7^{2^6} \equiv 9 \pmod{13}$

$7^{128} = 7^{2^7} \equiv 3 \pmod{13}$

$7^{256} = 7^{2^8} \equiv 9 \pmod{13}$

So the remainder when 7^{256} is divided by 13 is 9.

> The same answer can be obtained using $49 \equiv 10 \pmod{13}$, but using $-3 \pmod{13}$ makes the working easier.

Suppose we now want to find the remainder when 7^{15} is divided by 47.

We note that $7^2 \equiv 2 \pmod{47}$

$\Rightarrow 7^4 = (7^2)^2 \equiv 4 \pmod{47}$

$\Rightarrow 7^8 = (7^4)^2 \equiv 16 \pmod{47}$

Now we can stop because if we square again we obtain 7^{16}, and $16 > 15$.

We know that $15 = 8 + 4 + 2 + 1$

$$\Rightarrow 7^{15} = 7^{8+4+2+1} \equiv 7^8 \times 7^4 \times 7^2 \times 7 \pmod{47}$$
$$\equiv 16 \times 4 \times 2 \times 7 \pmod{47}$$
$$\equiv 64 \times 14 \pmod{47}$$
$$\equiv 17 \times 14 \pmod{47}$$
$$\equiv 238 \pmod{47}$$
$$\equiv 3 \pmod{47}$$

Hence the remainder when 7^{15} is divided by 47 is 3.

You would have noticed that in order to divide the exponent into powers of 2 you can convert it into a binary number, e.g. $15 = 1111_2 = 2^0 + 2^1 + 2^2 + 2^3 = 1 + 2 + 4 + 8$

We can write the method explained in the above example as an algorithm for computing $a^m \pmod{n}$:

1 Express m as a sum of powers of 2 (writing m in binary will help).

2 Compute $a^2 \pmod{n}$, $a^4 \pmod{n}$, $a^8 \pmod{n}$ ...

3 Combine the results in steps 1 and 2 to compute $a^m \pmod{n}$.

This method can be used whenever you want to compute remainders of large exponents, as in the next example.

Example 19

Find the value of $k \in \mathbb{Z}$, $0 \le k < 31$, such that $5^{51} \equiv k \pmod{31}$.	
$51 = 110011_2$ $= 2^5 + 2^4 + 2^1 + 2^0$ $= 32 + 16 + 2 + 1$	*Write 51 in binary.* *Write 51 as sum of powers of 2.*
$5^2 = 25 \equiv -6 \pmod{31}$ $5^4 \equiv 36 \pmod{31} \equiv 5 \pmod{31}$ $5^8 \equiv 25 \pmod{31} \equiv -6 \pmod{31}$ $5^{16} \equiv 36 \pmod{31} \equiv 5 \pmod{31}$ $5^{32} \equiv 25 \pmod{31} \equiv -6 \pmod{31}$	*Use Theorem 4 to compute remainders for powers of 5 up to 32.*
$5^{51} \equiv 5^{32} \times 5^{16} \times 5^2 \times 5 \pmod{31}$ $\equiv (-6) \times 5 \times (-6) \times 5 \pmod{31}$ $\equiv 36 \times 25 \pmod{31}$ $\equiv 5 \times (-6) \pmod{31}$ $\equiv -30 \pmod{31}$ $\equiv 1 \pmod{31}$ Therefore the remainder is 1.	*Combine results to find the required answer.*

In the next example we use the theorem to show that the specific difference of two numbers is a factor of a given number.

Example 20

Show that 39 divides $17^{48} - 5^{24}$.	
$17^{48} = 17^{32+16}$ $17^2 = 289 \equiv 16 \pmod{39}$ $17^4 \equiv 256 \pmod{39} \equiv -17 \pmod{39}$ $17^8 \equiv (-17)^2 \pmod{39} \equiv 16 \pmod{39}$ $17^{16} \equiv -17 \pmod{39}$ $17^{32} \equiv 16 \pmod{39}$	*Write 48 as sum of powers of 2.* *Use Theorem 4 to compute remainders for powers of 17 up to 32.*
$17^{48} = 17^{32+16} = 17^{32} \times 17^{16}$ $\equiv 16 \times (-17) \pmod{39}$ $\equiv -272 \pmod{39}$ $\equiv 1 \pmod{39}$	*Combine results to obtain $17^{48} \pmod{39}$.*
$5^{24} = 5^{16+8}$ $5^2 \equiv -14 \pmod{39}$ $5^4 \equiv 196 \pmod{39} \equiv 1 \pmod{39}$ $5^8 \equiv 1 \pmod{39}$ $5^{16} \equiv 1 \pmod{39}$	*Write 24 as a sum of powers of 2.* *Use Theorem 4 to compute remainders for powers of 5 up to 16.*
$5^{24} = 5^{16+8} \equiv 5^{16} \times 5^8 \equiv 1 \pmod{39}$ Since $17^{48} \equiv 5^{24} \pmod{39}$ it follows that $39 \mid 17^{48} - 5^{24}$.	*Combine results to obtain $5^{24} \pmod{39}$.*

Investigation

Copy and complete the following table. What do you notice?

k	$1^k \pmod 3$	$2^k \pmod 3$
1		
2		

Check whether the same is true in the following tables. Can you see any patterns?

k	$1^k \pmod 4$	$2^k \pmod 4$	$3^k \pmod 4$
1			
2			
3			

k	$1^k \pmod 5$	$2^k \pmod 5$	$3^k \pmod 5$	$4^k \pmod 5$
1				
2				
3				
4				

k	$1^k \pmod 6$	$2^k \pmod 6$	$3^k \pmod 6$	$4^k \pmod 6$	$5^k \pmod 6$
1					
2					
3					
4					
5					

k	$1^k \pmod 7$	$2^k \pmod 7$	$3^k \pmod 7$	$4^k \pmod 7$	$5^k \pmod 7$	$6^k \pmod 7$
1						
2						
3						
4						
5						
6						

You would have noticed many different patterns in each of these tables, particularly within columns. Can you see any patterns within rows?

Construct similar tables for (mod n) where $n = 8, 9, 10$ and 11.

Look at all the tables constructed and compare those rows which produce the same number.

Write a general rule for all those rows, e.g. in the first table you would notice that $2^k \equiv 1 \pmod 3$.

Can you make a conjecture?

? Pierre de Fermat was the son of a wealthy leather merchant in 17th century France. As a young man he attended the universities of Toulouse and Orléans. By the early 1630's he was a fully-fledged lawyer. He was a passionate mathematician and kept up his work in this field as a hobby. He tended to share his mathematical work with other mathematicians in France rather than publish them. He is renowned for his work known as "Fermat's Last Theorem" which states that the set of equations $a^n + b^n = c^n$ are insoluble if $n > 2$. He wrote this statement in the narrow margin of a book, along with the following statement: *"I have a truly marvelous demonstration of this proposition which this margin is too narrow to contain."* This hypothesis continued to baffle mathematicians for three whole centuries and although the proof was so elusive, it created many new ideas and discoveries in the field of mathematics. It was only in 1993, after a full decade of seclusion, that Andrew Wiles managed to work out a proof. The proof makes use of intricate modern mathematics and could not possibly be the same proof that Fermat alludes to in his copy of *Arithmetica*. We will never know whether Fermat had actually proved his proposition. If he had, it would not have been using the mathematics that Wiles did, thus the proof is lost forever. Fermat's Last Theorem is not to be confused with the Little Theorem which is explored next.

In the investigation on pages 67 and 68 you may have noticed the following results.

- $a^2 \equiv 1 \pmod 3$
- $a^4 \equiv 1 \pmod 5$
- $a^6 \equiv 1 \pmod 7$
- $a^{10} \equiv 1 \pmod{11}$

This leads to a conjecture for the next theorem.

> **Theorem 7: Fermat's Little Theorem**
>
> If p is prime and $\gcd(a, p) = 1$, then $a^{p-1} \equiv 1 \pmod p$.

Proof:

Consider the following multiples of a:

$a, 2a, 3a, \ldots, (p-1)a$

Suppose that in this list there are some integers $1 \le r < s \le p-1$ such that $ra \equiv sa \pmod p$.

Since $\gcd(a, p) = 1$ this means that $r \equiv s \pmod p$ cannot be true since $1 \le r < s \le p-1$. All the multiples in the list must be distinct and non-zero and therefore when we consider the remainders when each is divided by p they must give residues $1, 2, 3, \ldots, p-1$ but not necessarily in that order.

$\Rightarrow a \times 2a \times 3a \times \ldots \times (p-1)a \equiv 1 \times 2 \times 3 \times \ldots \times (p-1) \pmod p$

$\Rightarrow (p-1)! \times a^{p-1} \equiv (p-1)! \pmod p$

$\Rightarrow a^{p-1} \equiv 1 \pmod p$ Q.E.D.

An alternative but equivalent version of Fermat's Little Theorem states that $a^p \equiv a \pmod p$. It is left for you to show that the two versions are equivalent.
Although Fermat's Little Theorem applies only to primes, there are some composite integers, n, such that $a^{n-1} \equiv 1 \pmod n$. Such integers are called Pseudoprimes or Carmichael numbers. Examples of such numbers are $341 = 11 \times 31$ and $561 = 17 \times 33$.

The following examples illustrate how this theorem is used to compute powers modulo p.

Residue is another term for remainder. So given $a \equiv b \pmod n$, b is the residue when a is divided by n.

Example 21

Show that $5^{450} \equiv 1 \pmod{11}$.

$5^{540} = (5^{10})^{54}$ $5^{10} \equiv 1 \pmod{11} \Rightarrow (5^{10})^{54} \equiv 1 \pmod{11}$	*Notice that* $5^{540} = (5^{10})^{54}$. *Apply Fermat's Little Theorem.*

Example 22

Find the remainder when 512^{372} is divided by 13.	
$512 = 2^9$ and $372 = 31 \times 12$ $\Rightarrow (2^9)^{372} = 2^{9\times31\times12} = (2^{12})^{9\times31}$	*Reduce 512 into prime factors.*
$512^{372} = (2^{12})^{279} \equiv 1 \pmod{13}$	*Apply Fermat's Little Theorem.*

Example 23

Find the remainder when 3^{47} is divided by 23.	
$3^{47} \pmod{23}$ $\equiv 3^{44+3} \pmod{23}$ $\equiv 3^{44} \times 3^3 \pmod{23}$	*23 is a prime number so simplify the power to include 22.*
$\equiv (3^{22})^2 \pmod{23} \times 27 \pmod{23}$ $\equiv 1 \times 4 \pmod{23}$ $\Rightarrow$ the remainder when 3^{47} is divided by 23 is 4	*Apply Fermat's Little Theorem.*

Example 24

a Use Fermat's Little Theorem to compute $3^{302} \pmod 5$ and $3^{302} \pmod 7$. **b** Use your results and the Chinese Remainder Theorem to find $3^{302} \pmod{35}$.	
a $3^{302} = (3^4)^{75} \times 3^2$	*Rewrite 302 in terms of multiples of 4.*
$3^{302} \pmod 5$ $\equiv (3^4)^{75} \times 3^2 \pmod 5$ $\equiv 4 \pmod 5$	*Apply Fermat's Little Theorem and simplify.*
$3^{302} = (3^6)^{50} \times 3^2$	*Rewrite 302 in terms of multiples of 6.*
$3^{302} \pmod 7$ $\equiv (3^6)^{50} \times 3^2 \pmod 7$ $\equiv 2 \pmod 5$	*Apply Fermat's Little Theorem and simplify.*
b $x \equiv 4 \pmod 5$ $x \equiv 2 \pmod 7$ $M = 35$	*Write the two congruences with x and use the Chinese Remainder Theorem to solve.*
$7 \equiv 2 \pmod 5$ $49 \equiv 4 \pmod 5$ $5 \equiv 5 \pmod 7$ $\Rightarrow 15 \equiv 1 \pmod 7$ $\Rightarrow 30 \equiv 2 \pmod 7$ $49 + 30 = 79 \equiv 9 \pmod{35}$ Therefore $3^{302} \pmod{35} = 9$	*Multiply by 7 to obtain the required congruence*

Example 25

Use Fermat's Little Theorem to solve the linear congruence $3x \equiv 4 \pmod{11}$.	
$3x \equiv 4 \pmod{11}$ $3^{10}x \equiv 3^9 \times 4 \pmod{11}$ $\Rightarrow x \equiv (3^2)^4 \times 3 \times 4 \pmod{11}$ $\Rightarrow x \equiv (-2)^4 \times 1 \pmod{11}$ $\Rightarrow x \equiv 5 \pmod{11}$ General solution is $x = 11n + 5$.	*Since 11 is prime and* $\gcd(3, 11) = 1$ *multiply both sides by* 3^9. *Use Fermat's Little Theorem.*

Exercise 2E

1 **a** Find the value of $k \in \mathbb{Z}$, $0 \le k < 8$, such that $3^{65} \equiv k \pmod 8$.
b Find the value of $k \in \mathbb{Z}$, $0 \le k < 31$, such that $3^{1025} \equiv k \pmod{31}$.
c Find the value of $k \in \mathbb{Z}$, $0 \le k < 37$, such that $6^{543} \equiv k \pmod{37}$.

2 Find the remainder when 5^{62} is divided by 13.

3 Compute each of the following.
a $128^{350} \pmod{11}$
b $3444^{3233} \pmod{17}$

4 Use Fermat's Little Theorem to solve these linear congruences.
a $6x \equiv 5 \pmod{13}$
b $5x \equiv 2 \pmod 7$
c $7x \equiv 8 \pmod{11}$

5 Use Fermat's Little Theorem to find the remainder when $11^{158} + 4$ is divided by 13.

6 Show that 51 divides $13^{59} - 10^{68}$.

7 Show that if $23 \mid a^{110} - 1$ for all $a \in \mathbb{Z}^+$, $\gcd(a, 23) = 1$.

8 **a** Use Fermat's Little Theorem to compute $3^{2003} \pmod 5$, $3^{2003} \pmod 7$ and $3^{2003} \pmod{11}$.
b Use your results from part **a** and the Chinese Remainder Theorem to find $3^{2003} \pmod{385}$.

Review exercise

EXAM-STYLE QUESTIONS

1 **a** Which integers leave a remainder of 1 when divided by 2, and also a remainder of 1 when divided by 5?

b Find the integers which are divisible by 3 but leave a remainder of 1 when divided by 5.

2 **a** Show that 14 is a factor of $n^7 - n$ for all $n \in \mathbb{N}$.

b Use the result $2015 = 201 \times 10 + 5$ to show that $2^{2015} \equiv 10 \pmod{11}$.

c Find $2^{2015} \pmod 7$ and $2^{2015} \pmod{13}$.

3 **a** Use the Euclidean algorithm to express $\gcd(129, 1001)$ in the form $129m + 1001n$, where $m, n \in \mathbb{Z}$.

b Find the least positive solution of $129x \equiv 1 \pmod{1001}$.

c Find the general solution of $129x \equiv 1 \pmod{1001}$.

d Find the solution set of $129x \equiv 1 \pmod{1002}$.

4 **a** Find the general solution for the following system of congruences.

$x \equiv 1 \pmod 5$

$x \equiv 2 \pmod{11}$

$x \equiv 1 \pmod{13}$

b Find all the values of x such that $1000 < x < 3000$.

5 **a** Consider the integers $n = 1793$ and $m = 2981$, given in base 10.

i Express n and m in base 11.

ii Hence show that $\gcd(1793, 2981) = 11$.

b A list L is made up of $n + 1$ distinct positive integers. Prove that at least two members of L leave the same remainder on division by n.

c Consider these simultaneous equations:

$3x + y + 7z = a$

$2y + z = b$

$4x + 2z = c$

Show that $3a + b - c = 0 \pmod 5$.

6 **a** State two equivalent versions of Fermat's Little Theorem and show that they are equivalent.

b Hence or otherwise show that in base 10, the last digit of any integer n is always equal to the last digit of n^5.

c Show that this result is also true in base 15.

7 a Express 109 368 as a product of primes.

b Prove by induction that if N is an odd number, $N^n \equiv 1 \pmod 2$.

c Use the result of part **b** and Fermat's Little Theorem to show that $38 \mid (7^{19} + 31^{19})$.

8 Fermat's Little Theorem states that under certain conditions $a^p \equiv a \pmod p$.

a Show that this statement is equivalent to $a^{p-1} \equiv 1 \pmod p$.

b Show that this result is not true when $a = 3$, $p = 8$ and state which of the conditions is not satisfied.

c Find the smallest positive value of k satisfying the congruence $3^{32} \equiv k \pmod 8$.

9 a Given that $a \equiv b \pmod n$ and $c \equiv d \pmod n$, prove that $a + c \equiv b + d \pmod n$.

b Hence solve the system $\begin{cases} 13x + 8y \equiv 7 \pmod{15} \\ 5x + 22y \equiv 8 \pmod{15} \end{cases}$

c Show that $x^{41} - x + 2 \equiv 0 \pmod{41}$ has no solutions.

10 a State Fermat's Little Theorem.

b Given that p is an odd positive integer:

i Show that $\sum_{k=1}^{p} k^p \equiv 0 \pmod p$.

ii Given that $\sum_{k=1}^{p} k^{p-1} \equiv n \pmod p$ where $0 \le n \le p - 1$, find the value of n.

Chapter 2 summary

Definition: If $a \in \mathbb{Z}$ and $n \in \mathbb{Z}^+$, $n > 1$, then the **remainder** when a is divided by n is denoted by $r \equiv a(\text{mod } n)$.

Definition: If $a, b \in \mathbb{Z}$ and $n \in \mathbb{Z}^+$, $n > 1$, then a is **congruent** to b modulo n if n divides $a - b$. We use the following notation: $a \equiv b(\text{mod } n) \Leftrightarrow n \mid a - b$

Theorem 1:

Given $n \in \mathbb{Z}^+$, $a \equiv b(\text{mod } n) \Leftrightarrow a = b + kn$, where $k \in \mathbb{Z}$.

Theorem 2:

If $n \in \mathbb{Z}^+$ and a, b, c and $d \in \mathbb{Z}$, $a \equiv b(\text{mod } n)$ and $c \equiv d(\text{mod } n)$ then:

i $a + c \equiv b + d(\text{mod } n)$

ii $ac \equiv bd(\text{mod } n)$

Corollary:

i $a \equiv b(\text{mod } n) \Rightarrow ka \equiv kb(\text{mod } n)$

ii $a \equiv b(\text{mod } n) \Rightarrow a^k \equiv b^k(\text{mod } n)$

Definition: A congruence of the form $ax \equiv b(\text{mod } n)$ where $a, b \in \mathbb{Z}$, $n \in \mathbb{Z}^+$ and $x \in \mathbb{Z}$ is called a **linear congruence**.

Theorem 3:

If $a, n \in \mathbb{Z}^+$, $n > 1$ where a and n are relatively prime, it follows that an inverse of $a(\text{mod } n)$, denoted by $\bar{a}$, exists such that $\bar{a}a \equiv a\bar{a} \equiv 1(\text{mod } n)$, $\bar{a} \in \mathbb{Z}^+$, $\bar{a} < n$. Furthermore this inverse is unique modulo n.

Theorem 4:

For all $n, k \in \mathbb{Z}^+$, $n > k$, $na - k(\text{mod } n) = -k(\text{mod } n)$.

The Pigeonhole Principle:

If m pigeons occupy n pigeonholes and $m > n$, then at least one pigeonhole must be occupied by more than one pigeon.

The Generalized Pigeonhole Principle:

If n pigeonholes are occupied by $kn + 1$ or more pigeons, where $k \in \mathbb{Z}^+$, then at least one pigeonhole must be occupied by $k + 1$ or more pigeons.

Theorem 5: The Chinese Remainder Theorem

Given the system of linear congruences:

$x \equiv a_1 \pmod{m_1}$

$x \equiv a_2 \pmod{m_2}$

$x \equiv a_3 \pmod{m_3}$

$\vdots$

$x \equiv a_n \pmod{m_n}$

If $m_1, m_2, m_3, \ldots, m_n$ are pairwise relatively prime positive integers, there exists a unique solution modulo M where $M = m_1 \times m_2 \times m_3 \times \ldots \times m_n$.

In other words, there is a particular solution $0 \le x_0 < M$ and a general solution given by $x \equiv x_0 \pmod{m} + kM$.

Theorem 6:

Given that $a \equiv b \pmod{n}$ where $a, b \in \mathbb{Z}$ and $n \in \mathbb{Z}^+$ then $a^m \equiv b^m \pmod{n}$ for all $m \in \mathbb{Z}^+$.

Theorem 7: Fermat's Little Theorem

If p is prime and $\gcd(a, p) = 1$, then $a^{p-1} \equiv 1 \pmod{p}$.
An alternative equivalent version states that $a^p \equiv a \pmod{p}$.

3 Recursive patterns

CHAPTER OBJECTIVES:

10.11 Recurrence relations. Initial conditions, recursive definition of a sequence. Solution of first- and second-degree linear homogeneous recurrence relations with constant coefficients. The first degree linear recurrence relation $u_n = au_{n-1} + b$. Modelling with recurrence relations.

Before you start

You should know how to:

1 Find terms of a sequence defined recursively, e.g. the first four terms of the sequence defined by:

$$\begin{cases} u_0 = 1 \\ u_n = 2 + 3u_{n-1},\ n \in \mathbb{Z}^+ \end{cases}$$

are 1, 5, 17 and 53.

2 Recognize the first term and common difference of an arithmetic sequence given its general term, e.g. if $u_n = 2n + 5$ for $n \in \mathbb{N}$, the first term is 5 and the common difference is 2.

3 Recognize the first term and common ratio of a geometric sequence given its general term, e.g. if $u_n = 2^{2n+1}$ for $n \in \mathbb{Z}^+$, the first term is 8 and the common ratio is 4.

4 Find the sum S of a convergent geometric series, given its general term, e.g.

if $u_n = 3 \cdot \left(\frac{1}{4}\right)^{n+1}$ for $n \in \mathbb{Z}^+$ then

$$S = \frac{\frac{3}{16}}{1-\frac{1}{4}} = \frac{1}{4}$$

Skills check:

1 Find the first four terms of each sequence.

a $\begin{cases} u_0 = 2 \\ u_n = 5 - u_{n-1},\ n \in \mathbb{Z}^+ \end{cases}$

b $\begin{cases} u_0 = 4 \\ u_n = 2(u_{n-1} - 1),\ n \in \mathbb{Z}^+ \end{cases}$

2 State the first term and the common difference of these arithmetic sequences.

a $u_n = -2n + 1$ for $n \in \mathbb{N}$

b $u_n = n - 1$ for $n \in \mathbb{Z}^+$

3 State the first term and common ratio of the geometric sequence $u_n = 2 \cdot 3^{n+1}$ for $n \in \mathbb{Z}^+$.

4 Find the sum of the convergent geometric series defined by $u_n = 5 \cdot \left(\frac{1}{2}\right)^{2n+1}$ for $n \in \mathbb{Z}^+$.

Modelling and solving problems using sequences

Planning and predicting are crucial concepts when it comes to Business and Management. Discrete Mathematics provides useful tools to specialists in these areas, allowing them to create and analyze models and make informed decisions from them. In general, analyzing huge amounts of data and producing models is a difficult task even for the most able mathematicians specialized in this area. In this chapter we are going to look at situations that will give you an idea about the techniques involved and will allow you to appreciate the role of Discrete Mathematics in the development of other sciences, specifically Economics. These techniques will enable you to solve financial problems. These financial decisions are judgments you can make about your life and the future when you have the choice of investing in stocks, major appliances, a house, a car or starting a savings account. In the end they may determine how well you live!

Xunyu Zhou, developer of a rigorous mathematical basis for behavioural economics at Oxford said: *'Financial mathematics needs to tell not only what people ought to do, but also what people actually do. This gives rise to a whole new horizon for mathematical finance research: can we model and analyse the consistency and predictability in human flaws so that such flaws can be explained, avoided or even exploited for profit?'.*
(Quote from http://plus.maths.org/content/what-financial-mathematics)

3.1 Recurrence relations

As part of the core course you studied some recurrence relations when the recursive definition of a sequence was introduced. For example, you studied arithmetic and geometric sequences that are characterized by simple recurrence relations:

- If $\{u_n\}$ is an arithmetic sequence then it can always be defined by a recurrence relation of the form $\begin{cases} u_1 = a \\ u_n = u_{n-1} + d, n > 1 \end{cases}$ for constants $a, d \in \mathbb{R}$.
- If $\{u_n\}$ is a geometric sequence then it can always be defined by a recurrence relation of the form $\begin{cases} u_1 = a \\ u_n = u_{n-1} \times r, n > 1 \end{cases}$ for constants $a, r \in \mathbb{R}\setminus\{0\}, r \neq 1$.

In general, a recurrence relation establishes a clear rule relating terms of sequences, (usually) consecutive terms and includes an initial condition that allows you to calculate specific terms of the sequence.

For example, $\begin{cases} u_1 = 3 \\ u_n = u_{n-1} + 2, n > 1 \end{cases}$ defines the sequence 3, 5, 7, ...

In this case, the initial condition is $u_1 = 3$; the first term of the sequence is 3. For another example, the recurrence relation $\begin{cases} v_1 = 3 \\ v_n = 2v_{n-1}, n > 1 \end{cases}$ defines the sequence 3, 6, 12, 24, ..., but this time the sequence is geometric with first term 3 and common ratio 2. In both cases, you know how to find expressions for the general terms of these sequences: $u_n = 2n + 1$ and $v_n = 3 \times 2^{n-1}$, for $n \in \mathbb{Z}^+$.

However, it is not always straightforward to deduce a general formula for the nth term of a sequence as the following investigation shows you.

? Leonardo da Pisa (1175–1250) ala Fibonacci is known mostly because of the famous sequence that bears his name, and that is simply the solution of a problem that he included in his book *Liber Abaci*. However, the main contribution to mathematics of Leonardo da Pisa was the introduction of the Hindu-Arab number system in Europe.

Investigation on Fibonacci numbers

The Fibonacci numbers 1, 1, 2, 3, 5, 8, 13, 21, ... make up a famous mathematical pattern characterized by a simple recurrence relation:

$F_1 = 1$, $F_2 = 1$ and $F_n = F_{n-1} + F_{n-2}$, $n \in \mathbb{Z}^+$, $n > 2$.

Fibonacci numbers also provide a simple way of approximating Φ, the Golden Ratio:

Just take the quotient of two consecutive Fibonacci numbers:

$\frac{1}{1} = 1$; $\frac{2}{1} = 2$; $\frac{3}{2} = 1.5$; $\frac{5}{3} = 1.6\dot{6}\ldots$; $\frac{8}{5} = 1.6$; $\frac{13}{8} = 1.625$; ...

a Use a spreadsheet to generate the sequence of the quotients of consecutive Fibonacci numbers and calculate the relative error between each term and the Golden Ratio whose exact value is given by $\Phi = \frac{1+\sqrt{5}}{2}$.

b Repeat part **a** for other Fibonacci type sequences like 5, 3, 8, 11, 19, ... which differ from the original one in the choice of the first two terms.

c Using the fact that Φ is a solution of the equation $x^2 - x - 1 = 0$, show that $\Phi^2 = \Phi + 1$.

In mathematics you often deal with geometric sequences that are characterized by the recurrence relation $\frac{F_n}{F_{n-1}}$ = constant. You can generate as many geometric sequences as you wish: just choose the starting term and the constant (or common ratio) and start multiplying each term by the chosen constant! However, there is just one Fibonacci type sequence that is also a geometric sequence with positive terms. This is called the Golden sequence:

$1, \Phi, 1 + \Phi, 1 + 2\Phi, 2 + 3\Phi, 3 + 5\Phi, \ldots$

d Hence show that $\Phi^3 = \Phi(\Phi + 1) = 2\Phi + 1$.

e Explain why the Golden sequence is a geometric sequence with common ratio Φ, i.e. the Golden sequence can be written as $1, \Phi, \Phi^2, \Phi^3, \ldots$

There is another sequence with the same properties but its terms are alternately positive and negative: just replace Φ by the negative of its reciprocal $\Phi' = -\frac{1}{\Phi} = \frac{1-\sqrt{5}}{2}$ to obtain the reciprocal Golden sequence: $1, \Phi', 1 + \Phi', 1 + 2\Phi', 2 + 3\Phi', 3 + 5\Phi', \ldots$.

f Show that this sequence is also geometric with common ratio Φ', i.e. the reciprocal Golden sequence can be written as $1, \Phi', (\Phi')^2, (\Phi')^3, \ldots$

(*Hint: use the same method used in part* **c**, *as* $\Phi' = -\frac{1}{\Phi}$ *is also solution of* $x^2 - x - 1 = 0$.)

The Golden sequence and the reciprocal Golden sequence can also be seen as sequences of linear expressions in Φ whose coefficients of both the independent term and the term in Φ are exactly the Fibonacci numbers 0, 1, 1, 2, 3, 5, 8,.... (adding a starting term 0) as:

$0 + \Phi$, $1 + \Phi$, $1 + 2\Phi$, $2 + 3\Phi$, ... and this means that

$F_{n-1} + F_n\Phi = \Phi^n$ and $F_{n-1} + F_n\Phi' = (\Phi')^n$, $n \geq 1$.

g Solve both equations simultaneously to obtain a general formula for Fibonacci numbers $F_n = \frac{\Phi^n - (\Phi')^n}{\sqrt{5}} = \frac{\Phi^n - (-\Phi)^{-n}}{\sqrt{5}}$ known as Binet's formula.

This formula was derived by Jacques Binet in 1843, although the result was known to Euler, Daniel Bernoulli, and De Moivre more than a century earlier.

The investigation on Fibonacci numbers dealt with a second-degree recurrence relation, $u_{n+1} = u_n + u_{n-1}$, as it involves three consecutive terms of the sequence. To determine the solution – the Fibonacci sequence – you needed two initial conditions, for example $u_1 = u_2 = 1$, or $u_0 = 0$ and $u_1 = 1$.

Definition

A **recurrence relation** of order k is an equation that defines each further term of a sequence as a function of k preceding terms, i.e. $u_n = f(u_{n-1},\dots,u_{n-k})$.

Recurrence relations of order k are also called recurrence relations of degree k.

In general, as you will see in the next sections, the number of initial conditions to determine the general term of a sequence defined by a recurrence relation is given by its degree. So far you have worked with arithmetic and geometric sequences defined by first degree recurrence relations and one initial condition.

Example 1 shows you that sometimes more complicated recurrence relations represent, in fact, simple sequences. In this example we also deduce a general expression for the solution of the recurrence relation and prove it using the principle of strong mathematical induction you studied in Chapter 1.

Example 1

Consider a sequence $\{u_n\}$ defined by the recurrence relation $u_n = 3u_{n-2} + 2u_{n-1}$, $n \in \mathbb{Z}^+$, $n \geq 3$, with $u_1 = -1$ and $u_2 = 1$.

a Determine the first six terms of the sequence.

b Hence, conjecture a formula for the general term of the sequence.

c Use the principle of strong mathematical induction to prove your conjecture.

---

a $-1, 1, -1, 1, -1, 1$

1.1 Example 1

	A n	B un	C	D
◆				
2	2	1		
3	3	-1		
4	4	1		
5	5	-1		
6	6	1		

B3 =3·b1+2·b2

You may use a GDC spreadsheet to find the terms of the sequence.

b $u_n = (-1)^n$, $n \in \mathbb{Z}^+$

Notice that this sequence is a geometric sequence with first term and common ratio both equal to −1.

c Let P_n: $u_n = (-1)^n$, $n \in \mathbb{Z}^+$.

State the claim that you want to prove.

When $n = 1$,

$u_1 = (-1)^1 \Rightarrow u_1 = -1$ ∴ true.

When $n = 2$,

$u_2 = (-1)^2 \Rightarrow u_2 = 1$ ∴ true.

Therefore P_1 and P_2 are true.

Prove that the claim is true for the two initial values. We need both values since this is a recurrence relation involving two previous terms.

Assume that P_i is true for all $1 \leq i \leq k$.

When $n = k + 1$, $k > 1$.

$$\begin{aligned} u_{k+1} &= 3u_{k-1} + 2u_k \\ &= 3 \times (-1)^{k-1} + 2 \times (-1)^k \\ &= (-1)^{k-1} \times (3 + 2 \times (-1)) \\ &= (-1)^{k-1} \times \underbrace{(-1)^2}_{1} \\ &= (-1)^{k-1+2} \\ &= (-1)^{k+1} \end{aligned}$$

Make the assumption (inductive step). Prove that the statement is true for $n = k + 1$.

Therefore P_n is true for $n = k + 1$.

Since P_1 and P_2 are true and it was shown that given P_i is true for all $1 \leq i \leq k$, P_{k+1} is also true, by strong mathematical induction it follows that P_n is true for all $n \in \mathbb{Z}^+$.

Example 2 is about a recurrence relation whose solution is the difference between two geometric sequences. To prove it we will again use the principle of strong mathematical induction.

Example 2

Show that if $u_n = 5u_{n-1} - 6u_{n-2}$, $n \geq 2$, with $u_1 = 1$ and $u_2 = 5$, then for all $n \in \mathbb{Z}^+$, $u_n = 3^n - 2^n$.

Proof:	
$P_n: u_n = 3^n - 2^n$	*State the claim you want to prove.*
When $n = 1$, $3^1 - 2^1 = 1 = u_1$ When $n = 2$, $3^2 - 2^2 = 5 = u_2$ Therefore P_1 and P_2 are true.	*Prove that the claim is true for the two initial values. We need both values since this is a recurrence relation involving the two previous terms.*
Assume that P_n is true for all $1 \leq i \leq k$.	*Make the assumption (inductive step).*
When $n = k + 1$, $k > 1$ $u_{k+1} = 5u_k - 6u_{k-1}$ $= 5(3^k - 2^k) - 6(3^{k-1} - 2^{k-1})$ $= 5 \times 3^k - 5 \times 2^k - 2 \times 3^k + 3 \times 2^k$ $= 3 \times 3^k - 2 \times 2^k$ $= 3^{k+1} - 2^{k+1}$ Therefore P_n is true for $n = k + 1$.	*Prove that the statement is true for $n = k + 1$.*
Since P_1 and P_2 are true and it was shown that given P_n is true for all $2 \leq n \leq k$, P_{k+1} is also true, it follows by the principle of strong mathematical induction that P_n is true for all $n \in \mathbb{Z}^+$.	

Exercise 3A

1 Use the principle of strong mathematical induction to prove that given $u_1 = 3$, $u_2 = 5$ and, for all $n \geq 2$, $u_{n+1} = 3u_n - 2u_{n-1}$, then $u_n = 2^n + 1$.

2 Consider the recurrence relation defined by $a_1 = 2$ and, for all $n \geq 1$, $a_{n+1} = a_n^2$.

- **a** Find the first 5 terms of the sequence.
- **b** Hence, conjecture a formula for the general term of the sequence.
- **c** Use the principle of strong mathematical induction to prove your conjecture.

3 Use the principle of strong mathematical induction to show that given $a_1 = 900$, $a_2 = 1780$ and, for all $n \geq 2$, $a_{n+1} = 2a_n - a_{n-1} + 10{\cdot}2^n$, then $a_n = 840n + 20 + 20{\cdot}2^n$.

Questions **4** to **6** involve the Fibonacci sequence defined by

$$F_0 = 0,\ F_1 = 1, \text{ and } F_n = F_{n-1} + F_{n-2} \text{ for } n \in \mathbb{Z}^+,\ n \geq 2$$

4 Show that $\sum_{i=0}^{n} F_i = F_{n+2} - 1$, for all $n \in \mathbb{Z}^+$.

5 **a** Find the values of F_n for $2 \leq n \leq 8$.

b Show that for $1 \leq n \leq 5$, $\sum_{i=0}^{n} F_i^2 = F_n \times F_{n+1}$.

c Use the principle of strong mathematical induction to prove that for all $n \in \mathbb{Z}^+$, $\sum_{i=0}^{n} F_i^2 = F_n \times F_{n+1}$.

6 Use Binet's formula $F_n = \dfrac{\Phi^n - (\Phi')^n}{\sqrt{5}} = \dfrac{\Phi^n - (-\Phi)^{-n}}{\sqrt{5}}$ to show that

a $F_n \times F_{n+2} = F_{n+1}^2 - (-1)^n$ **b** $\lim_{n \to +\infty} \dfrac{F_{n+1}}{F_n} = \Phi$

Although you may be asked to conjecture general term formulas for any recurrence relation and provide proof using strong mathematical induction, only for special first- and second-degree recurrence relations are you required to use systematic methods to deduce these solutions. For this reason, in the following sections we will cover some types recurrence relations of first- and second-degree.

3.2 Solution of first-degree linear recurrence relations and applications to counting problems

In this section you are going to learn a method to find the general term of a sequence that satisfies a recurrence relation of the form $u_n = a \cdot u_{n-1} + f(n)$ where $a \in \mathbb{R}\backslash\{0\}$, $n \in \mathbb{Z}^+$ and f is a function of n of the form $f(n) = b + cn$, for $b, c \in \{\mathbb{R}\}\backslash\{0\}$.
These recurrence relations can be classified into:

i homogeneous first-degree recurrence relations if $b = c = 0$

ii inhomogeneous first-degree recurrence relations otherwise.

Some texts may refer to inhomogeneous equations as nonhomogeneous. It is also usual to refer to first-degree recurrence relations as first-order recurrence relations.

Let's look at case **i** first, since you have already dealt with these types of sequences: they are geometric sequences. Let's deduce again their general term and focus on the process because we will need to apply it to case **ii**:

$$u_n = a \cdot u_{n-1} = a{\cdot}(a \cdot u_{n-2}) = a{\cdot}(a{\cdot}(a \cdot u_{n-2})) = \ldots = a^{n-1} \cdot u_1$$

which means that the general term is $u_n = a^{n-1} \cdot u_1$.

We proceed now to the general case **ii** but this time we are going to simplify the notation to make the pattern clear:

$u_n = a \cdot u_{n-1} + f(n)$, i.e.

$u_n = a \cdot (a \cdot u_{n-2} + f(n-1)) + f(n) = a^2 \cdot u_{n-2} + a \cdot f(n-1) + f(n)$

Proceeding with the same method:

$u_n = a^3 \cdot u_{n-3} + a^2 \cdot f(n-2) + a \cdot f(n-1) + f(n)$

Eventually, if we continue the process we will obtain

$$u_n = a^{n-1} \cdot u_1 + a^{n-2} \cdot f(2) + a^{n-3} \cdot f(3) + \ldots + a \cdot f(n-1) + f(n)$$

or, using sigma notation,

$$u_n = a^{n-1} \cdot u_1 + \sum_{k=0}^{n-2} a^k \cdot f(n-k)$$

If the initial condition is given in terms of u_0, then

$$u_n = a^n \cdot u_0 + a^{n-1} \cdot f(1) + a^{n-2} \cdot f(2) + \ldots + a \cdot f(n-1) + f(n)$$

or, using sigma notation,

$$u_n = a^n \cdot u_0 + \sum_{k=0}^{n-1} a^k \cdot f(n-k)$$

This formula shows that the solution of the recurrence relation depends on a summation problem. This is a difficult problem and its general scope goes beyond the requirements of the Mathematics HL syllabus. We will therefore focus instead on particular cases and explore their applications in solving counting problems.

Depending on the initial condition given we may need to relate u_n with u_1 or with u_0.

Case 1: $f(n) = b$, i.e. $u_n = a \cdot u_{n-1} + b$.

The general solution is $u_n = a^{n-1} \cdot u_1 + b \cdot (1 + a + a^2 + \ldots + a^{n-2})$. Using the formula for the sum of consecutive terms of a geometric sequence, we obtain $u_n = a^{n-1} \cdot u_1 + b \cdot \dfrac{a^{n-1} - 1}{a - 1}$.

The solution for Case 1 is of the form $u_n = A \cdot a^{n-1} + B$. To find the general solution, we can use the recurrence relation, calculate the value of u_2 and then use the values of the first two terms to determine the values of A and B.

Example 3 shows you three methods that you can use to solve this type of first-degree recurrence relation.

This alternative method to solve Case 1 can be used to solve recurrence relations when you know the form of the solution and just need to find the values of some parameters using the initial conditions given.

Example 3

Solve the recurrence relation $u_n = 2u_{n-1} + 1$ with initial condition $u_1 = 1$.

Method I

$$u_n = 2 \cdot u_{n-1} + 1$$
$$= 2 \cdot (2 \cdot u_{n-2} + 1) + 1 = 2^2 u_{n-2} + 2 + 1$$
$$= 2^2(2 \cdot u_{n-3} + 1) + 2 + 1 = 2^3 \cdot u_{n-3} + 2^2 + 2 + 1$$
$$\dots$$
$$= 2^{n-1} \cdot u_1 + \underbrace{2^{n-2} + \dots + 2^2 + 2 + 1}_{\text{sum of } n-1 \text{ consecutive terms of a geometric progression}}$$
$$= 2^{n-1} + 1 \cdot \frac{2^{n-1} - 1}{2 - 1} = 2 \cdot 2^{n-1} - 1$$
$$= 2^n - 1$$

Apply the recurrence relation $n - 2$ times to obtain an expression for u_n in terms of the first term.

Use $u_1 = 1$ and the formula for sum of consecutive terms of a geometric progression with first term 1 and common ratio 2 to obtain the general formula required.

Method II

The solution is:

$$u_n = 2^{n-1} \cdot 1 + 1 \cdot \frac{2^{n-1} - 1}{2 - 1}$$
$$= 2^{n-1} + 2^{n-1} - 1$$
$$= 2 \times 2^{n-1} - 1 \Rightarrow u_n = 2^n - 1$$

Use $u_n = a^{n-1} \cdot u_1 + b \cdot \dfrac{a^{n-1} - 1}{a - 1}$ where $a = 2$, $b = 1$ and simplify the expression.

Method III

The solution is:

$$u_n = A \cdot 2^{n-1} + B, \; A, B \in \mathbb{R}.$$
$$u_2 = 2 \cdot u_1 + 1 \Rightarrow u_2 = 3 \text{ since } u_1 = 1$$
$$\begin{cases} 1 = A \cdot 2^0 + B \\ 3 = A \cdot 2^1 + B \end{cases} \Rightarrow \begin{cases} A = 2 \\ B = -1 \end{cases}$$
$$\therefore u_n = 2^n - 1.$$

The solution is of the form $u_n = A \cdot a^{n-1} + B$ with $a = 2$.
Use the recurrence relation to calculate u_2.
Use $u_1 = 1$ and $u_2 = 3$ to determine the values of A and B.

The recurrence relation in Example 3 is the solution of a famous counting problem: the n-ring Tower of Hanoi puzzle.

According to the legend of the Tower of Hanoi, temple priests are to transfer a tower consisting of 64 fragile disks of gold from one part of the temple to another, one disk at a time. The disks are arranged in order, no two of them the same size, with the largest on the bottom and the smallest on top. Because of their fragility, a larger disk may never be placed atop a smaller one, and there is only one intermediate location where disks can be temporarily placed. The legend says that before the priests complete their task the temple will turn into dust and the world will end. How long do they have to complete the task if it takes them one minute to transfer each disk?

Investigation

Try the Tower of Hanoi puzzle and convince yourself that the recurrence relation $u_{n+1} = 2u_n + 1$ with $u_1 = 1$ provides a model to solve this problem.

Case 2: $a = 1$

As $a = 1$ and $f(n) = b + cn$, the general solution $u_n = a^n \cdot u_0 + \sum_{k=0}^{n-1} a^k \cdot f(n-k)$ can be simplified to $u_n = u_0 + \sum_{k=0}^{n-1}(c \cdot (n-k) + b) \Rightarrow u_n = u_0 + \sum_{k=0}^{n-1}(b + cn - ck)$. This means that we can find the general term of the sequence $\{u_n\}$ simply by applying our knowledge of arithmetic progressions, as shown in Examples 4 to 6.

Example 4

Solve the recurrence relation $u_n = u_{n-1} + n$ with initial condition $u_0 = 1$.

Method I

The solution is $u_n = 1 + \sum_{k=0}^{n-1}(n-k)$:

Use $u_n = u_0 + \sum_{k=0}^{n-1}(c \cdot (n-k) + b)$ where $b = 0$, $c = 1$.

$\Rightarrow u_n = 1 + n \cdot \frac{n+1}{2} = \frac{n^2+n+2}{2}$

Use $S_n = \frac{a_0 + a_{n-1}}{2} \cdot n$ where $a_k = n - k$.

Method II

$u_n = u_{n-1} + n$

Use the recurrence relation to explore the pattern and obtain a formula for u_n in terms of n.

$= u_{n-2} + (n-1) + n$

$= u_{n-3} + (n-2) + (n-1) + n$

Note that you obtain an arithmetic series with n terms, first term 1 and last term n.

$= \underbrace{u_0}_{1} + \underbrace{1 + \ldots + (n-2) + (n-1) + n}_{\frac{n(n+1)}{2}}$

$\therefore u_n = \frac{n^2+n+2}{2}$

When you conjecture a formula like in Method II, you may be asked to prove it using strong mathematical induction.

The recurrence relation in Example 4 is a special one as it models another well-known counting problem:

Suppose that we draw n straight lines on a piece of paper so that every pair of lines intersects, but no three lines intersect at a common point. Into how many regions do these n lines divide the plane?

Let's investigate this problem by examining the situation for small values of n.

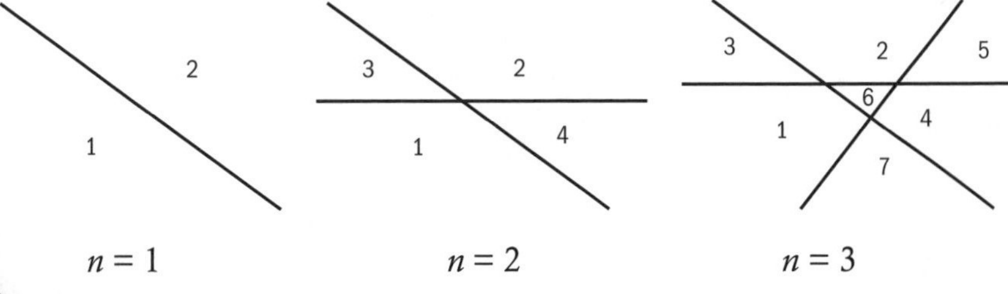

Investigation

Explore the intersecting lines problem using dynamic geometry software and convince yourself that this problem is modelled by the recurrence relation $u_{n+1} = u_n + n$ with $u_0 = 1$.

Extension: General case $u_n = a \cdot u_{n-1} + b + cn$ when $a \neq 0$ and $c \neq 0$

We saw that the solution is of the form $u_n = a^n \cdot u_0 + \sum_{k=0}^{n-1} a^k \cdot (b + c(n-k))$ but it is very difficult to obtain an explicit expression for the general term unless you are guided or provided with extra information that allows you to determine a simplified expression for such a sum.

An alternative approach to this type of problem and a more general method is the following:

Rewrite the recurrence relation in the form $u_n - a \cdot u_{n-1} = b + cn$ and consider the auxiliary recurrence relation $v_n - a \cdot v_{n-1} = 0$ (i.e. $v_n = a \cdot v_{n-1}$). This homogeneous recurrence relation has general solution $v_n = k \cdot a^n$ as we showed before. The general solution of the recurrence relation $u_n = a \cdot u_{n-1} + b + cn$ is of the form $u_n = k \cdot a^n + An + B$ where k, A and B are constants to be determined.

An auxiliary recurrence relation is a homogeneous recurrence relation that we use to tackle problems involving more complex recurrence relations.

The next example shows you how to apply this alternative method.

Example 5

Consider the recurrence relation $u_n = 2u_{n-1} + n + 1$, $n \geq 1$ with $u_0 = -1$.

a Write down the general solution of the corresponding homogeneous equation $v_n = 2v_{n-1}$.

b Determine the values of A and B in $p_n = An + B$ such that the expression is a solution of the recurrence relation $u_n = 2u_{n-1} + n + 1$.

c Hence write down the solution of the recurrence relation $u_n = 2u_{n-1} + n + 1$, $n \geq 1$ with $u_0 = -1$.

---

a The solution is $v_n = k \cdot 2^n$, $k \in \mathbb{R}$.

The solution of $v_n = a \cdot v_{n-1}$ is $v_n = k \cdot a^n$.

b $p_n = An + B \Rightarrow p_{n-1} = A(n-1) + B$

$p_n = p_{n-1} + n + 1$

$An + B = 2(A(n-1) + B) + n + 1$

$An + B = (2A + 1)n + (-2A + 2B + 1)$

$\therefore A = -1$ and $B = -3$

$p_n = -n - 3$

The general solution of the recurrence relation $u_n - a \cdot u_{n-1} = b + cn$ is of the form $u_n = k \cdot a^n + An + B$ where k, A and B are constants to be determined by substituting $p_n = An + B$ and $p_{n-1} = A(n-1) + B$ into the recurrence relation.

c The general solution is $u_n = k \cdot 2^n - n - 3$.

$u_0 = -1 \Rightarrow -1 = k \cdot 2^0 - 0 - 3 \Rightarrow k = 2$

$\therefore u_n = 2^{n+1} - n - 3$

Use initial condition to determine the value of k.

Analyze Example 5 carefully. Note that the solution of the inhomogeneous recurrence relation $u_n - a \cdot u_{n-1} = b + cn$ is obtained by determining the solution $v_n = k \cdot a^n$ of the corresponding homogeneous recurrence relation $v_n - a \cdot v_{n-1} = 0$ and then adding a particular solution $p_n = An + B$ where A and B are constants that can determined by substituting $p_n = An + B$ into the recurrence relation $u_n - a \cdot u_{n-1} = b + cn$. Therefore the solution of this inhomogeneous recurrence relation is of the form $u_n = \underbrace{k \cdot a^n}_{v_n} + \underbrace{An + B}_{p_n}$

The next example shows you how to apply this method efficiently.

Example 6

Solve the recurrence relation $u_n = 3u_{n-1} + 4n + 2$, $n \geq 1$, with initial condition $u_0 = 1$.	
Consider the auxiliary homogeneous recurrence relation $v_n = 3 \cdot v_{n-1}$ whose solution is $v_n = k \cdot 3^n$. Consider $p_n = An + B$.	*$v_n = k \cdot a^n$ is the solution of the corresponding homogeneous recurrence relation $v_n = a \cdot v_{n-1}$*
$An + B = 3 \cdot (A(n-1) + B) + 4n + 2$ $An + B = (3A + 4)\, n + 3B - 3A + 2$ $\therefore\ A = 3A + 4 \Rightarrow A = -2$ and	*Substitute $p_n = An + B$ into $u_n = 3u_{n-1} + 4n + 2$ to determine the values of A and B.*
$B = 3B + 3A + 2 \Rightarrow B = -4$	*The solution is the form*
Therefore the solution is of the form $u_n = k \cdot 3^n - 2n - 4$ $u_0 = 1 \Rightarrow 1 = k \cdot 3^0 - 2 \times 0 - 4 \Rightarrow k = 5$ $u_n = 5 \cdot 3^n - 2n - 4$	$u_n = \underbrace{k \cdot a^n}_{v_n} + \underbrace{An + B}_{p_n}$ *Use the initial condition to find the value of k.*

Note that you can solve inhomogeneous recurrence relations of the form $u_n = a \cdot u_{n-1} + f(n)$ using the method shown in the previous examples only when $f(n) = b + cn$; for other inhomogeneous recurrence relations where f is not a linear function of n, the form of the particular solution needs to be given.

Exercise 3B

1 Solve the recurrence relations.

a $u_n = 3u_{n-1} - 2$ with $u_1 = 2$

b $u_n = 2u_{n-1} - 1$ with $u_1 = 3$

c $u_n = -u_{n-1} + 2$ with $u_0 = 2$

d $u_n = -2u_{n-1} + 3$ with $u_0 = -1$

2 Solve the recurrence relations.

a $u_n = 2u_{n-1} + 3n + 1$ with $u_1 = 1$

b $u_n = 3u_{n-1} + n + 1$ with $u_1 = 2$

EXTENSION QUESTIONS

Questions 3-8 require you to combine knowledge of different topics and follow the guidance given to find the solution of the recurrence relations. These type of recurrence relations are not explicitly mentioned in the syllabus and you are not required to learn any general method to tackle them.

3 Consider the recurrence relation $u_n = u_{n-1} + 2n - 1$ with initial condition $u_0 = 1$.

a Show that $\sum_{k=0}^{n-1}(2(n-k)-1) = n^2, n \in \mathbb{Z}^+$.

b Find the general solution of the recurrence relation.

c Hence find an expression for u_{n-1}.

d Verify that your answer to part **b** is correct by substituting the expressions for u_n and u_{n-1} into the recurrence relation given.

4 a Use the principle of mathematical induction to show that $\sum_{k=1}^{n} k^2 = \frac{n(n+1)(2n+1)}{6}$.

b Hence, solve the recurrence relation $u_n = u_{n-1} + 3n^2$ with $u_1 = 10$.

5 Use the substitution $v_n = u_n^2$ to solve the recurrence relation $u_n^2 = u_{n-1}^2 + 1$ with $u_1 = 1$.

6 Given n distinct objects, let p_n denote the number of possible arrangements (or permutations) of these objects if displayed in a row.

a Show that this $\{p_n\}$ satisfies the recurrence relation

$p_{n+1} = (n+1)p_n$ with $p_1 = 1$.

b Hence show that $p_n = n!$

$\{p_n\}$ is an example of a recurrence relation of order 1 with variable coefficients.

7 Solve the recurrence relation $u_n = 2u_{n-1} + 3n^2$, $n \geq 0$ with initial condition $u_0 = 1$, given that a particular solution of the equation is of the form $p_n = An^2 + Bn + C$.

8 Solve the recurrence relation $u_n = 5u_{n-1} + 3^n$, $n \geq 0$ with initial condition $u_0 = -1$, given that a particular solution of the equation is of the form $p_n = A \cdot 3^n$.

3.3 Modelling with first-degree recurrence relations

Financial problems

Companies advertising loans and investment products try to make their products look as attractive as possible. Also, they often have different ways of calculating the interest, and the products might involve different investment periods. It is important that you are informed and use your knowledge of mathematics before making any important decision such as:

- agreeing to the amount of each regular payment for a given loan and the number of years over which the loan is to be repaid
- deciding how much money to invest right now in return for specific cash amounts to be received in the future

- calculating the amount of the pension fund required on the date of retirement to give a fixed income every year for a certain number of years
- determining the fair market value of a bond.

A bond is a certificate issued by a government or a public company promising to repay borrowed money at a fixed rate of interest at a specified time.

Loans and amortizations

Arrangements involving savings and loans often involve making a regular payment at fixed intervals. For example, a savings account might involve saving a certain amount of money every month for a number of years. A *mortgage* might involve borrowing a certain amount of money and repaying it in equal instalments over time. Calculations involving such regular payment schedules, when they are considered in terms of the present values of the payments, as in loans, or the future values, as in investments, will involve the summation of a geometric series, as shown in the next example.

Some of these problems can be solved using your previous knowledge of core topics. However, here we are developing systematic and efficient approaches that allow you to deal with these types of problem with less guidance.

Example 7

Zixian borrows €10 000 at an interest rate of 6% p.a.(per annum). He wants to repay the loan in five equal instalments over five years, with the first payment one year after he takes out the loan. How much should each payment be?

Let each payment equal A. Note that the present value of each payment A_n made after n years is given by the recurrence relation $A_n = \frac{A_{n-1}}{1.06}$ with $A_1 = \frac{A}{1.06}$.	*When regular payments are used to pay off a loan, then we are interested in calculating their present values.* *6% p.a. means 6% per year which results in the factor* $1+\frac{6}{100} = 1.06$.
$\frac{A}{1.06} + \frac{A}{1.06^2} + \ldots + \frac{A}{1.06^5} = 10000$	*The total of all the payments is equal to the loan amount.*
$\frac{\frac{1}{1.06}\left(1-\frac{1}{1.06^5}\right)}{\left(1-\frac{1}{1.06}\right)} = 10000 \Rightarrow 4.212363\ldots A = 10000$ (6 dp)	*This is a geometric series with* $n = 5$, *first term* $a = \frac{A}{1.06}$ *and* $r = \frac{1}{1.06}$.
$\therefore A = \frac{10000}{4.212363} \Rightarrow A = 2373.96 = €2373.96$	

If we switch to months as the unit of time, then we first have to determine what monthly interest rate, compounded monthly, is equivalent to the interest rate of 6% per annum. This involves getting the twelfth root of 1.06, which is 1.004867551.

We can treat the present values as a geometric series with first term $a = \frac{A}{1.004867551}$,

common ratio $r = \frac{1}{1.004867551}$ and number of terms $n = 60$.

$$\frac{\frac{A}{1.004867551}\left(1-\left(\frac{1}{1.004867551}\right)^{60}\right)}{\left(1-\frac{1}{1.004867551}\right)} = 10000 \Rightarrow 51.923A = 10000$$

This gives a monthly repayment of $A =$ €192.59.

You may want to calculate the amount Zixian can save over 5 years if he makes loan payments monthly rather than yearly.

This type of problem is so common that it is convenient to derive a formula to shortcut the calculation for the regular repayment A. By considering the general case of an amortized loan with interest rate I, taken out over t years, for a loan amount of P, a geometric series can be used to derive the general formula: $A = P\frac{I(1+I)^t}{(1+I)^t - 1}$

You will derive this formula later as an exercise.

If you use this amortization formula with $I = 0.004867551$, $P = 10\,000$, and $t = 60$ you obtain

$$A = 10000 \cdot \frac{0.004867551(1.004867551)^{60}}{(1.004867551)^{60} - 1} = €192.5898\ldots = €192.59 \text{ (2 dp)}$$

Investments and compound interest

When regular payments are being used for investment, we are interested in their future values since this tells us how much we can expect to have when the investment matures. A scheduled regular payment over time will again give rise to a geometric series as shown in the next example.

Example 8

A local bank is offering a regular monthly savings account with a fixed interest rate of 4.00% p.a. on balances up to €15 000. If Kristian saves €100 per month, starting today and assuming the rate stays the same, how much will he have in his account in five years' time?

$100(1.04)^{\frac{60}{12}} + 100(1.04)^{\frac{59}{12}} + \ldots + 100(1.04)^{\frac{2}{12}} + 100(1.04)^{\frac{1}{12}}$ $= \frac{100(1.04)^5\left(1-\left(\frac{1}{1.04^{\frac{1}{12}}}\right)^{60}\right)}{\left(1-\frac{1}{1.04^{\frac{1}{12}}}\right)} = €6639.57$	*The amount saved each month has a future value given by the recurrence relation* $A_n = 1.04^{-\frac{1}{12}}A_{n-1}$ *with* $A_1 = 100 \times 1.04^{\frac{60}{12}}$ *This is a geometric series with first term* $100(1.04)^5$*, common ratio* $100(1.04)^{-\frac{1}{12}}$ *and 60 terms.*

Games and probability problems

First-degree recurrence relations are also a good tool to deal with endless games like the next example shows you.

Example 9

Ewa and Ericka are playing a game where the first player has some advantage over the other player:

- The probability that the first player wins a round is $\frac{3}{5}$;
- The probability that the second player wins a round is $\frac{2}{5}$.

Today Ewa plays first and will continue playing first until she loses, at which point she needs to pass the first turn to Ericka. Let w_n denote the probability that Ewa wins the nth round.

a State the value of w_1 and find the value of w_2.

b Show that $\{w_n\}$ satisfies the recurrence relation $w_n = \frac{1}{5}w_{n-1} + \frac{2}{5}$.

c Solve this recurrence relation to find the probability that Ewa wins the third game.

d Hence show that if Ewa and Ericka play this game for a long time, the probability that Ewa wins the nth game is approximately 0.5. Comment on the significance of this result.

a $w_1 = \frac{3}{5}$

$$w_2 = \frac{3}{5} \times \frac{3}{5} + \frac{2}{5} \times \frac{2}{5} = \frac{13}{25}$$

Ewa plays first and the probability that the first player wins a round is $\frac{3}{5}$;

$$w_2 = \frac{3}{5}P\text{ (won 1st round)} + \frac{2}{5}P\text{ (lost 1st round)}$$

b $w_n = \frac{3}{5}w_{n-1} + \frac{2}{5}(1 - w_{n-1}) = \frac{1}{5}w_{n-1} + \frac{2}{5}$

When Ewa is the first player, the probability that she wins a round is $\frac{3}{5}$; when she is the second player (because she lost the previous round) the probability is $\frac{2}{5}$.

c $$w_n = \frac{3}{5}\left(\frac{1}{5}\right)^{n-1} + \frac{2}{5}\cdot\frac{\left(\frac{1}{5}\right)^{n-1} - 1}{\frac{1}{5} - 1}$$

$$= \frac{3}{5}\left(\frac{1}{5}\right)^{n-1} + \frac{1}{2}\left(1 - \left(\frac{1}{5}\right)^{n-1}\right) = \frac{1}{10}\left(\frac{1}{5}\right)^{n-1} + \frac{1}{2}$$

Use $u_n = a^{n-1} \cdot u_1 + b \cdot \frac{a^{n-1} - 1}{a - 1}$ where $a = \frac{1}{5}$ and $b = \frac{2}{5}$; simplify the expression.

d $$\lim_{n\to\infty} w_n = \lim_{n\to\infty}\left(\frac{1}{10}\left(\frac{1}{5}\right)^{n-1} + \frac{1}{2}\right) = \frac{1}{2}$$

Therefore Ewa and Ericka have approximately the same probability of winning the nth round if they play long enough.

Calculate $\lim_{n\to\infty} w_n$ and compare with $1 - \lim_{n\to\infty} w_n$.

Exercise 3C

1 Dispanshu won a prize and has been given a choice of two options:

A: Receive a payment of €1500 at the beginning of each month for 25 years.
B: Take a lump sum L instead.

Dispanshu decides to take option **B**.

a Find the minimum value of the lump sum L he should accept assuming an interest rate of 4% p.a., given that the interest is paid monthly.

b If Dispanshu invests the lump sum he receives for 20 years in an account that pays 4% p.a., determine the final value of Dispanshu's investment after the 20 years.

2 Timur opens an account with an interest rate of 3.4% p.a. He deposits €1000 at the beginning of each year for 10 years. Find the value in ten years of Timur's annuity. How much will he earn on his investment?

3 Nischay plans to deposit €400 at the beginning of each month for 10 years in an account earning 3.25% p.a. Find the lump sum of money Nischay would need to invest now to achieve the same final value as Timur from question **2**.

4 The management company of Zikuan's apartment block estimates that they will need €30 000 in 4.5 years' time to repaint the outside of the building and common areas.
If regular payments are made to a fund earning 2.75% p.a.,

a Calculate the interest rate per month, if paid and compounded monthly, that would be equivalent to an effective annual rate of 2.75%.

b Find the amount that needs to be deposited at the beginning of each month to meet this target.

c Find the total interest that will be earned in the 4.5 years.

5 Ewa and Ericka are playing another game where the first player also has some advantage over the other player:

- The probability that the first player wins a round is $\frac{4}{7}$;
- The probability that the second player wins a round is $\frac{3}{7}$.

Today Ericka plays first and will continue playing first until she loses, at which point she needs to pass the first turn to Ewa.
Let $\{w_n\}$ denote the probability that Ericka wins the nth round.

a State the value of w_1 and find the value of w_2.

b Show that $\{w_n\}$ satisfies the recurrence relation $w_n = \frac{1}{7}w_{n-1} + \frac{3}{7}$.

c Solve this recurrence relation to find the probability that Ericka wins the fifth round.

d Find the probability that Ericka wins the nth round as n gets very large.

3.4 Second-degree linear homogeneous recurrence relations with constant coefficients

In this section you are going to learn a method to find the general term of a sequence that satisfies a recurrence relation of the form $a \cdot u_{n+2} + b \cdot u_{n+1} + c \cdot u_n = 0$ where $a, b, c \in \mathbb{R}$, $a \neq 0$ and $n \in \mathbb{N}$ (or sometimes $n \in \mathbb{Z}^+$). These recurrence relations are called **second-degree linear homogeneous recurrence relations** and to determine a solution we need two initial conditions.

These types of recurrence relations are very useful to model a large class of counting problems, including some famous ones like the Fibonacci problem, described by the recurrence relation $F_n = F_{n-1} + F_{n-2}$, with initial conditions $F_1 = 1$, $F_2 = 1$.

? The original problem that Fibonacci investigated in the year 1202 was about how fast rabbits could breed in ideal circumstances. This was the ideal scenario: suppose a newly-born pair of rabbits, one male, one female, are put in a field. Rabbits are able to mate at the age of one month so that at the end of its second month a female can produce another pair of rabbits. We also suppose that our rabbits never die and that the female always produces one new pair (one male, one female) every month from the second month on. The puzzle that Fibonacci posed was to determine the number of rabbit pairs will there be after one year.

Although these recurrence relations seem more complicated, they are in fact easy to solve using a systematic method:

Step 1: Express the second-degree linear homogeneous recurrence relation given in the form $a \cdot u_{n+2} + b \cdot u_{n+1} + c \cdot u_n = 0$ and write down its **auxiliary equation** $ax^2 + bx + c = 0$.

Step 2: Solve the quadratic equation obtained using an appropriate method.

Step 3: The general term of your solution depends on the number and type of solutions of the auxiliary equation. In each case, A and B are constants to be determined using the initial conditions provided.

Case 1: Two distinct real roots α_1 and α_2. The general solution is $u_n = A\alpha_1^n + B\alpha_2^n$.

Case 2: A double real root α_1. The general solution is $u_n = A\alpha_1^n + Bn\alpha_1^n$.

Case 3: Two conjugate complex roots $r\text{cis}(\pm\theta)$. The general solution is $u_n = r^n(A \cos n\theta + B \sin n\theta)$.

The auxiliary or characteristic equation of a second-degree recurrence relation is a quadratic equation associated to the recurrence relation by the coefficients *a*, *b* and *c* of its terms.

Examples 10 to 12 show you how to apply the formula in each of the cases.

Example 10

Solve the recurrence relation $u_n = u_{n-1} + u_{n-2}$ with initial conditions $u_0 = 0$ and $u_1 = 1$.	
$u_{n+2} = u_{n+1} + u_n \Rightarrow u_{n+2} - u_{n+1} - u_n = 0$ The auxiliary equation is $x^2 - x - 1 = 0$	*Re-arrange the equation to the form* $a \cdot u_{n+2} + b \cdot u_{n+1} + c \cdot u_n = 0$ *and write down the equation* $ax^2 + bx + c = 0$.
and its solutions are $x = \frac{1 \pm \sqrt{5}}{2}$.	*Solve the quadratic equation:* $x = \frac{-b \pm \sqrt{b^2 - 4ac}}{2a}$
The general term is of the form $u_n = A\left(\frac{1+\sqrt{5}}{2}\right)^n + B\left(\frac{1-\sqrt{5}}{2}\right)^n$	*The general solution is* $u_n = A\alpha_1^n + B\alpha_1^n$.
$\begin{cases} u_0 = 0 \\ u_1 = 1 \end{cases} \Rightarrow \begin{cases} A + B = 0 \\ A(1 + \sqrt{5}) + B(1 - \sqrt{5}) = 2 \end{cases}$	*Substitute* $n = 0$ *and* $n = 1$.
$\Rightarrow \begin{cases} A + B = 0 \\ (A + B) + (A - B)\sqrt{5} = 2 \end{cases}$	*Solve simultaneously for A and B.*
$\therefore A = \frac{1}{\sqrt{5}}$ and $B = -\frac{1}{\sqrt{5}}$	
$\therefore u_n = \frac{1}{\sqrt{5}}\left(\frac{1+\sqrt{5}}{2}\right)^n - \frac{1}{\sqrt{5}}\left(\frac{1-\sqrt{5}}{2}\right)^n$	*The solution is the famous Binet's general formula for Fibonacci numbers* F_n *that you had already deduced in the investigation on pages 79–80.*

Example 11

Solve the recurrence relation $u_{n+2} = 2u_{n+1} - u_n$ with initial conditions $u_0 = 1$ and $u_1 = 2$.	
$u_{n+2} = 2u_{n+1} - u_n \Rightarrow u_{n+2} - 2u_{n+1} + u_n = 0$	*Re-arrange the equation to the form* $a \cdot u_{n+2} + b \cdot u_{n+1} + c \cdot u_n = 0$ *and write down the equation* $ax^2 + bx + c = 0$.
The auxiliary equation is $x^2 - 2x + 1 = 0$ and its solution is $x = 1$. The general term is of the form $u_n = A + Bn$	*Solve the quadratic equation by inspection.* *The general solution is* $u_n = A\alpha^n + Bn\alpha^n$.
$\begin{cases} u_0 = 1 \\ u_1 = 2 \end{cases} \Rightarrow \begin{cases} A = 1 \\ A + B = 2 \end{cases}$	*Substitute* $n = 0$ *and* $n = 1$. *Solve simultaneously for A and B.*
$\therefore A = B = 1$ $\therefore u_n = n + 1$	

Example 12

Solve the recurrence relation $u_{n+2} + u_{n+1} + u_n = 0$ with initial conditions $u_0 = u_1 = 1$.	
The auxiliary equation is $x^2 + x + 1 = 0$ and its solutions are $x = \frac{-1 \pm i\sqrt{3}}{2}$	*Write down the equation $ax^2 + bx + c = 0$. Solve the quadratic to obtain two conjugate complex roots.*
i.e. $x = \text{cis}\left(\pm\frac{2\pi}{3}\right)$	*Write them in the form $r\text{cis}(\pm\theta)$.*
The general term is of the form $u_n = A\cos\frac{2n\pi}{3} + B\sin\frac{2n\pi}{3}$	*The general solution is $u_n = r^n(A\cos n\theta + B\sin n\theta)$.*
$\begin{cases} u_0 = 1 \\ u_1 = 1 \end{cases} \Rightarrow \begin{cases} A = 1 \\ \cos\frac{2\pi}{3} + B\sin\frac{2\pi}{3} = 1 \end{cases}$	*Substitute $u_1 = u_0 = 1$.*
$\therefore A = 1,\ B = \sqrt{3}$	*Solve simultaneously for A and B.*
$\therefore u_n = \cos\frac{2n\pi}{3} + \sqrt{3}\sin\frac{2n\pi}{3}$	

Examples 13 and 14 show you real life applications of second-degree homogeneous recurrence relations.

Example 13

The number of cases of student sickness due to a very contagious virus at Gauss International School is growing in such a way that the growth rate in any week is twice the growth rate during the previous week. Suppose that during the first week when students started getting sick the number of cases of infection was 20, and in the second week 25 cases were identified. Let $\{S_n\}$ denote the number of students sick due to this virus during week n.

a Show that $\{S_n\}$ is defined by the recurrence relation $S_{n+2} - S_{n+1} = 2(S_{n+1} - S_n)$ and state appropriate initial conditions.

b Solve the recurrence relation to obtain an expression for S_n in terms of n.

c Find S_7.

d State a limitation of this model in the real life context provided.

a $\underbrace{S_{n+2} - S_{n+1}}_{\text{increase during week } n+1} = 2(\underbrace{S_{n+1} - S_n}_{\text{increase during week } n})$	*The difference between consecutive terms of $\{S_n\}$ represents the number of new cases of this virus.*
Initial conditions: $S_1 = 20$ and $S_2 = 25$	*In the first two weeks there were 20 and 25 cases registered, respectively.*

b $S_{n+2} - 3S_{n+1} + 2S_n = 0$ $x^2 - 3x + 2 = 0 \Rightarrow (x-1)(x-2) = 0$ $\therefore x = 1$ or $x = 2$ The general solution is $\therefore S_n = A + B \cdot 2n$ $S_1 = 20 \Rightarrow A + 2B = 20$ $S_2 = 25 \Rightarrow A + 4B = 25$ $\therefore A = 15$ and $B = 2.5$ $\therefore S_n = 15 + 2.5 \times 2^n$	*Re-arrange the equation into the form* $a \cdot u_{n+2} + b \cdot u_{n+1} + c \cdot u_n = 0$ *and write down its auxiliary equation* $ax^2 + bx + c = 0$ *and solve it.* *As there are two distinct real roots,* α_1 *and* α_2*, the general solution is* $u_n = A\alpha_1^n + B\alpha_1^n$. *Use initial conditions to determine the values of A and B.* *Substitute the values into the expression to obtain the solution to the problem.*
c $S_7 = 15 + 2.5 \times 2^7 = 335$ **d** The model is valid for just a few values of n as S_n cannot exceed the total number of students in the school.	*Substitute n by 7.* *For example, consider limitations related to restrictions of values of n in relation to size of the school.*

Example 14

In a colony of birds there were initially 12 pairs (12 male and 12 female). Assume that none of the birds produced eggs in their first year but in subsequent years each pair produced 4 eggs (2 male, 2 female) and no birds died. Show that the colony's population can be described by the recurrence relation $u_n = u_{n-1} + 2u_{n-2}$ for $n \geq 2$ with $u_0 = u_1 = 12$, where u_n is the number of pairs of birds at the beginning of the nth year. Hence find the minimum number of years until the population exceeds 200 pairs of birds.

From the second year on, the number of pairs is given by the sum of the number of pairs in the previous year plus the newborn birds which are exactly twice as many as the number of pairs two years before.	*u_n is the number of pairs of birds at the beginning of the nth year.* *None of the birds produced eggs in their first year.*
$u_n = u_{n-1} + 2u_{n-2} \Rightarrow u_n - u_{n-1} - 2u_{n-2} = 0$ $x^2 - x - 2 = 0 \Rightarrow (x+1)(x-2) = 0$ $\therefore x = -1$ or $x = 2$ The general solution is $\therefore u_n = A(-1)^n + B \cdot 2^n$ $u_1 = 12 \Rightarrow -A + 2B = 12$ $u_2 = 12 \Rightarrow A + 4B = 12$ $\therefore A = -4$ and $B = 4$ $\therefore u_n = -4 \Rightarrow (-1)^n + 2^{n+2}$ $u_n > 200 \Rightarrow -4(-1)^n + 2^{n+2} > 200$ $\therefore n \geq 5$ The population will exceed 200 pairs in 5 years.	*Re-arrange the equation into the form* $a \cdot u_{n+2} + b \cdot u_{n+1} + c \cdot u_n = 0$ *and write down its auxiliary equation* $ax^2 + bx + c = 0$ *and solve it.* *As there are two distinct real roots,* α_1 *and* α_2*, the general solution is* $u_n = A\alpha_1^n + B\alpha_1^n$. *Use initial conditions to determine the values of A and B.* *Use a GDC to solve the inequality and find the minimum value of n that satisfies the condition.*

1.1 Example 14

	A n	B un	C	D
◆				
1	1	12		
2	2	12		
3	3	36		
4	4	84		
5	5	204		

B5 =b3+2·b4

Exercise 3D

EXAM-STYLE QUESTIONS

1 Solve the second-degree recurrence relations.

a $u_n = 3u_{n-1} + 4u_{n-2}$ with $u_0 = u_1 = 1$

b $u_n = u_{n-2}$ with $u_0 = 3$ and $u_1 = 2$

c $u_n = 4u_{n-1} - 4u_{n-2}$ with $u_0 = 1$ and $u_1 = 3$

d $u_n - 2u_{n-1} + 5u_{n-2} = 0$ with $u_0 = u_1 = 1$

2 For each of the sequences $\{u_n\}$,

i write down a recurrence relation of the form $u_{n+2} - u_{n+1} = k(u_{n+1} - u_n)$ where $k \in \mathbb{R}, k \neq 0$.

ii re-arrange the equation into the form $a \cdot u_{n+2} + b \cdot u_{n+1} + c \cdot u_n = 0$ and solve it to obtain a general term for $\{u_n\}$.

a u_n: 3, 6, 12, 24, 48, ...

b u_n: 1, 2, 6, 22, 44, 88, ...

3 In an experiment, the pressure of gas in a closed container is measured in regular intervals. Let P_n represent the pressure (in standard units) at the nth time it is measured. Given that $P_1 = 8$, $P_2 = 6$, and for $n > 2$, $P_n = \frac{P_{n-1} + P_{n-2}}{2}$, find a formula for P_n in terms of n. Hence, state the value of the pressure of the gas after a long period of time in the container.

4 Consider the following counting problem:
Viktoria can climb one or two steps at a time when going up the school stairs. Let V_n denote the number of distinct ways to get to the nth step.

a Explain why $V_1 = 1$, $V_2 = 2$ and $V_n = V_{n-1} + V_{n-2}$, for $n \geq 3$.

b State the relation between the sequence $\{V_n\}$ and the Fibonacci numbers.

c Hence state an expression for the general term of $\{V_n\}$.

5 Show the following property of Fibonacci numbers:

$$\sum_{i=0}^{n} F_{2i} = F_{2n+1} - 1$$

a using strong mathematical induction

b using Binet's formula.

6 State clearly the recurrence relation and the initial conditions that define the Lucas numbers (see box at the right). Hence find a generating expression of this sequence $\{L_n\}$.

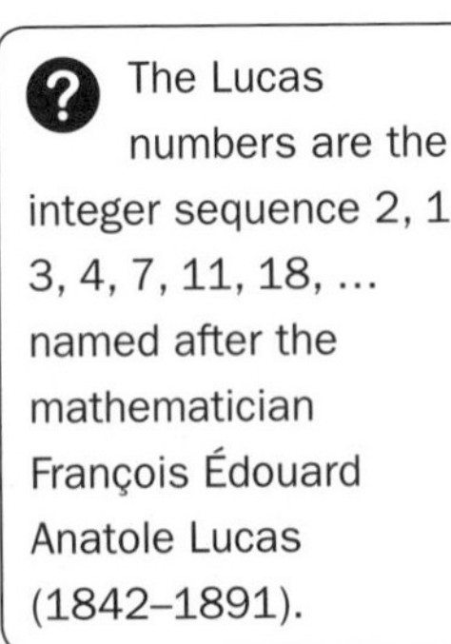

The Lucas numbers are the integer sequence 2, 1, 3, 4, 7, 11, 18, ... named after the mathematician François Édouard Anatole Lucas (1842–1891).

7 Use the general expressions from the sequences $\{F_n\}$ and $\{L_n\}$ to show the following properties of the Lucas and Fibonacci numbers.

a $L_n = F_{n-1} + F_{n+1}$

b $5 \cdot F_n = L_{n-1} + L_{n+1}$

c $F_{2n} = F_n \cdot L_n$

Review exercise

EXAM-STYLE QUESTIONS

1 Solve the following recurrence relations.

a $u_n = -3u_{n-1} + 1$ with $u_0 = 1$

b $u_n = 5u_{n-1} - 2$ with $u_0 = 6$

c $u_n = 6u_{n-1} - 5n + 1$ with $u_0 = 10$

d $u_n = 7u_{n-1} - 8n$ with $u_0 = -3$

2 The sequence $\{u_n\}$, $n \in \mathbb{Z}^+$, $n \geq 2$, satisfies the second-degree recurrence relation $u_{n+1} = 7u_n - 12u_{n-1}$.
Given that $u_0 = -2$ and $u_1 = 12$, use the principle of strong mathematical induction to show that $u_n = -20 \cdot 3^n + 18 \cdot 4^n$.

3 The sequence $\{u_n\}$, $n \in \mathbb{Z}^+$, $n \geq 2$, satisfies the second-degree recurrence relation $u_n = -4(u_{n-1} + u_{n-2})$.
Given that $u_1 = 1$ and $u_2 = 8$, use the principle of strong mathematical induction to show that $u_n = (-2)^{n-1}(5n - 6)$.

4 The recurrence relation $u_n + A \cdot u_{n-1} + B \cdot u_{n-2} = C$, $n \geq 2$, defines the sequence whose first five terms are 0, 2, 5, 9 and 12.

a Determine the values of A, B and C.

b Hence find the next 3 terms of the sequence $\{u_n\}$.

5 Find an expression for the general solution of the recurrence relation $u_{n+2} + 9u_n = 0$. Hence find the solution that verifies the initial conditions $u_0 = -4$ and $u_1 = 2$.

6 A bank pays 8% interest p.a. for a long-term investment. Elias decides to invest €1000 right away; Nischay decides that instead he will open an account with just €100, and then deposit €100 at the beginning of each year.

a Write down a homogeneous recurrence relation whose general solution a_n represents the total amount Elias has in his bank account at the end of the nth year.

b Write down a recurrence relation whose general solution b_n represents the total amount Nischay has in his bank account at the end of the nth year, should he follow his savings plan.

c Solve both recurrence relations (note that $a_0 = 1000$ and $b_0 = 100$).

d Hence find the minimum number of years Nischay needs to save until his savings exceed the amount Elias has as a result of his initial investment of €1000.

7 Solve the recurrence relation $u_{n+2} - 3u_n = 0$, with $u_0 = 2$ and $u_1 = 6$.

8 Find the general solution of the recurrence relation $v_{n+2} = v_{n-1} + 2v_n$, $n \in \mathbb{N}$, given that $v_0 = 1$ and $v_1 = 3$.

9 Find the general solution of the recurrence relation $u_{n+2} = 5u_{n-1} - 6u_n$, $n \in \mathbb{Z}^+$. State the initial conditions that result in a general solution $u_n = 3^n$, $n \in \mathbb{Z}^+$.

10 Consider the two recurrence relations $u_n = 5u_{n-1} + 2v_{n-1}$ and $v_n = u_{n-1} + 2v_{n-1}$, with $u_0 = 2$ and $v_0 = 1$.

a By eliminating v_n and v_{n-1}, show that $u_n = 7u_{n-1} - 8u_{n-2}$.

b Hence solve the recurrence relations and find expressions for u_n and v_n.

Chapter 3 summary

Definition: A recurrence relation of degree k is an equation that defines each further term of a sequence as a function of k preceding terms, i.e. $u_n = f(u_{n-1}, \ldots, u_{n-k})$.

A **first-degree recurrence relation** is of the form $u_n = a \cdot u_{n-1} + f(n)$ where $a \in \mathbb{R} \setminus \{0\}$, $n \in \mathbb{Z}^+$ and f is a function of n of the form $f(n) = b + cn$, for $b, c \in \{\mathbb{R}\}$.

These recurrence relations can be classified into:

i homogeneous first-degree recurrence relations if $b = c = 0$;

ii inhomogeneous (or non-homogeneous) first-degree recurrence relations otherwise.

General solutions of first-degree recurrence relations:

If $u_n = a \cdot u_{n-1} + b$, the general solution is $u_n = a^{n-1} \cdot u_1 + b \cdot \dfrac{a^{n-1} - 1}{a - 1}$

or $u_n = a^n \cdot u_0 + b \cdot \dfrac{a^n - 1}{a - 1}$

If $u_n = u_{n-1} + b + c \cdot n$, the general solution is $u_n = u_0 + \sum_{k=0}^{n-1} (c \cdot (n - k) + b)$.

If $u_n = a \cdot u_{n-1} + b + c \cdot n$, the general solution is of the form $u_n = a^n \cdot u_0 + An + B$ where A and B are constants to be determined by substituting $p_n = An + B$ into $u_n = a \cdot u_{n-1} + b + c \cdot n$.

A **second-degree recurrence relation** is of the form $a \cdot u_{n+2} + b \cdot u_{n+1} + c \cdot u_n = 0$ where $a, b, c \in \mathbb{R}$, $a \neq 0$ and $n \in \mathbb{N}$ (or sometimes $n \in \mathbb{Z}^+$).

General solutions of second-degree homogeneous recurrence relations

Step 1: Express the second-degree linear homogeneous recurrence relation given in the form $a \cdot u_{n+2} + b \cdot u_{n+1} + c \cdot u_n = 0$ and write down its **auxiliary equation** $ax^2 + bx + c = 0$.

Step 2: Solve the quadratic equation obtained using an appropriate method.

Step 3: The general term of your solution depends on the number and type of solutions of the auxiliary equation.
In each case, A and B are constants to be determined using the initial conditions provided.

Case 1: Two distinct real roots α_1 and α_2:
The general solution is $u_n = A\alpha_1^n + B\alpha_2^n$.

Case 2: A double real root α_1:
The general solution is $u_n = A\alpha_1^n + Bn\alpha_1^n$.

Case 3: Two conjugate complex roots $r\text{cis}(\pm\theta)$:
The general solution is $u_n = r^n(A \cos n\theta + B \sin n\theta)$.

The **Fibonacci sequence** is defined by $F_0 = 0$, $F_1 = 1$ and $F_n = F_{n-1} + F_{n-2}$, for $n \in \mathbb{Z}^+$, $n \geq 2$.

Binet's formula for Fibonacci numbers: $F_n = \frac{1}{\sqrt{5}}\left(\frac{1+\sqrt{5}}{2}\right)^n - \frac{1}{\sqrt{5}}\left(\frac{1-\sqrt{5}}{2}\right)$.

Converting miles into kilometres

As $\lim_{n\to+\infty} \frac{F_{n+1}}{F_n} = F$, the ratio of two consecutive numbers tends to the Golden Ratio as numbers get bigger and bigger. The Golden Ratio is a number and it happens to be approximately 1.618. Coincidentally, there are 1.609 kilometers in a mile, which is very close to the value of the Golden Ratio. This means that if you take two consecutive terms F_n and F_{n+1} in the Fibonacci sequence, for each F_n number of miles the conversion to kilometres is approximately equal to F_{n+1}. The accuracy of this conversion depends on the order of the terms used as for small values of n the values of the ratio $\frac{F_{n+1}}{F_n}$ vary; however as n gets larger the ratio quickly tends to the Golden Ratio as you can verify with your graphical calculator:

Golden ratio I

	A fn	B ratio	C
◆			
4	3.	1.5	
5	5.	1.6666667	
6	8.	1.6	
7	13.	1.625	
8	21.	1.6153846	

B ratio

Golden ratio II

	A fn	B ratio	C
◆			
9	34.	1.6190476	
10	55.	1.6176471	
11	89.	1.6181818	
12	144.	1.6179775	
13	233.	1.6180556	

B ratio

4 From folk puzzles to a new branch of mathematics

CHAPTER OBJECTIVES:

10.7 Graphs, vertices, edges, faces. Adjacent vertices, adjacent edges. Degree of a vertex, degree sequence. Handshaking lemma. Simple graphs; connected graphs; complete graphs; bipartite graphs; planar graphs; trees; weighted graphs, including tabular information. Subgraphs; complements of graphs. Euler's relation: $v - e + f = 2$ (including proof); theorems for planar graphs including $e \leq 3v - 6$, $e \leq 2v - 4$, leading to the results that κ_5 and $\kappa_{3,3}$ are not planar.

10.8 Walks, trails, paths, circuits, cycles. Eulerian trails and circuits. Hamiltonian paths and cycles

Before you start

You should know how to:

1 Know and use set notation and terminology, e.g. given the universal set $U = \{1, 2, 3, 4, 5, 6\}$ and its subsets $A = \{1, 2, 3\}$ and $B = \{1, 3, 5\}$, the complement of A is $A' = \{4, 5, 6\}$, the intersection of A and B is $A \cap B = \{3\}$ and their union is $A \cup B = \{1, 2, 3, 5\}$.

2 Recognize regular solids and their elements, e.g. a cube is a regular solid with 6 faces, 8 vertices and 12 edges.

3 Know how to apply the Pigeonhole Principle to solve counting problems, e.g. If you pick five cards from a standard deck of 52 playing cards, then at least two cards will be of the same suit because each of the five cards can belong to one of four suits. By the Pigeonhole Principle, two or more must belong to the same suit.

Skills check:

1 Given the universal set $U = \{1, 2, 3, 4, 5, 6, 7, 8\}$, and its subsets $A = \{2, 4, 6\}$ and $B = \{5, 6, 7\}$, find:

a $A \cap B$ **b** $A \cup B$

c A' **d** $(A \cap B)'$

2 Sketch a quadrangular pyramid and state its number of vertices, edges and faces.

3 In an experiment, scientists want to find two people with the same blood group (A, B, O or AB). In order to save time, the blood samples will be collected and processed simultaneously. Find the smallest number of samples that should be collected.

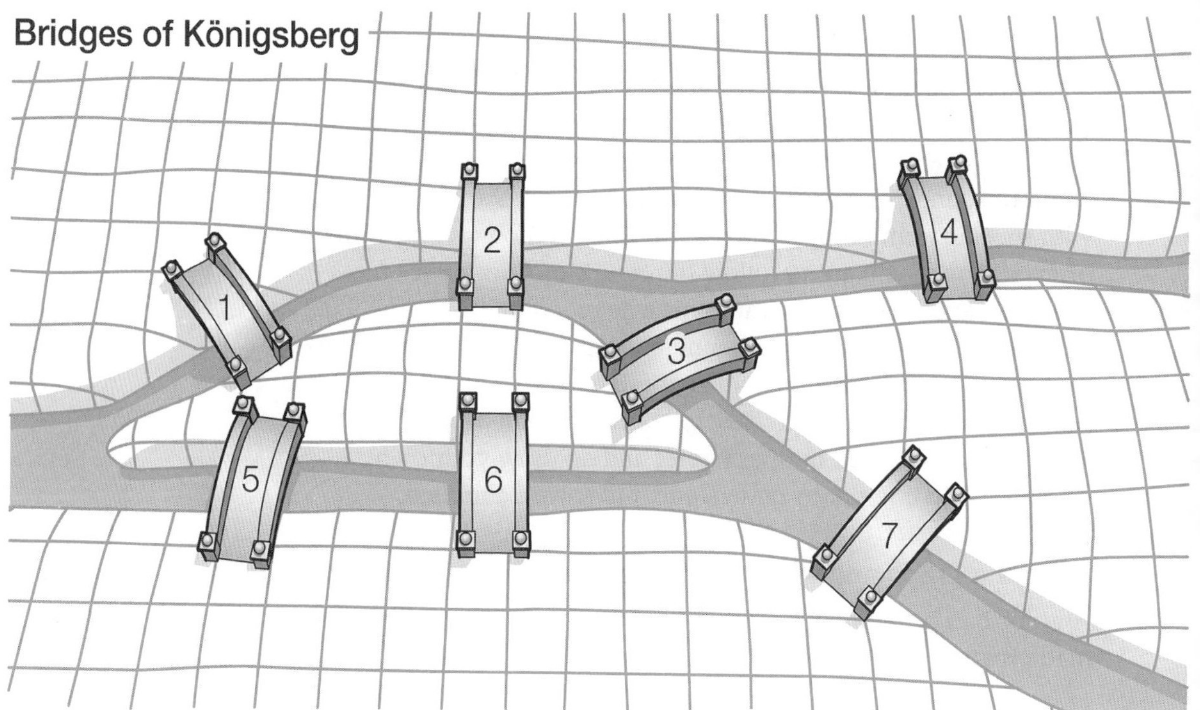

Introduction to Graph Theory

In the history of mathematics the first reference to Graph Theory is associated with a popular problem – the bridges of Königsberg – solved by Leonhard Euler almost 300 years ago.

This branch of mathematics then grew in sometimes rather informal and disorganized ways, as many times the methods were devised by curious mathematicians who were challenged by rather recreational problems. In fact, many of these problems and puzzles intrigued mathematicians exactly because their statements were so simple but required rather clever methods to tackle them. With the advent of computers, and as the importance of systematic and clear approaches to problem solving became crucial, Graph Theory evolved to a recognized branch of mathematics, characterized by its specific terminology and methods, and valued for its wide range of applications to contemporary problems.

The Königsberg bridge problem asks if the seven bridges of the city of Königsberg (now Kaliningrad, in Russia), over the river Preger can all be traversed in a single trip without going back over any of the bridges already crossed, with the additional requirement that the trip ends in the same place it began. In 1736 Euler proved that this cannot be done, and his proof is considered the beginning of Graph Theory.

4.1 Terminology and classification of graphs

In this section we are going to define the most common terms used in Graph Theory and look at key examples to give you a clear idea about the concepts involved. This branch of mathematics is laden with specific terminology and you must learn the exact meaning of each term if you are to follow the results of the course. Precision and rigour are paramount when it comes to Graph Theory. On the other hand, most terms and concepts can be described by diagrams – Graph Theory is all about diagrams! – and this makes some concepts evident and easy to learn. So, look carefully at all diagrams provided in this chapter and do draw your own whenever you attempt to understand and solve a problem.

What is a graph and what are its elements?

A graph can be defined in terms of a very simple concept in mathematics: the concept of set.

Definition

A graph $G = (V, E)$ consists of two sets:

1 A set V of vertices (also called points or nodes)

2 A set E of unordered pairs of vertices called edges (also called arcs)

In a graph, two vertices A and B are said to be **adjacent** if they are joined by an edge, i.e. if there is an edge $a = \text{AB}$.

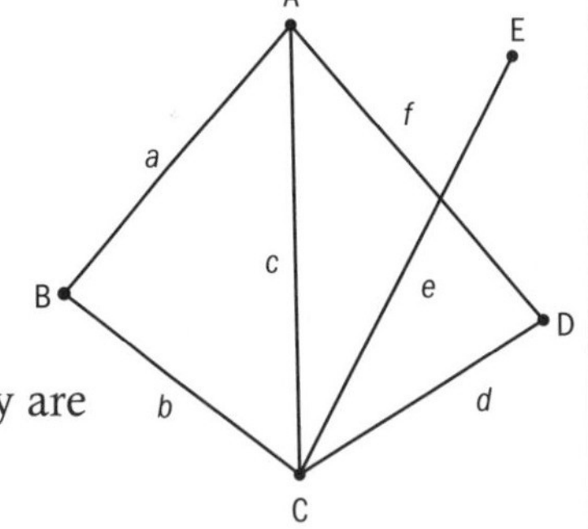

In this case, vertices A and B are said to be **incident** on edge *a*. Two edges are adjacent if they have a common vertex, i.e. there is a vertex incident with both edges. For example, the edges *a* and *b* are adjacent because the vertex B is incident with both of them.

The edges of the graph can also be labeled *a, b, c, d, e and f* as shown in the diagram.

Definition

The **order** and **size** of a graph G are defined in terms of the number of elements of the sets V and E: $v = |V|$ = order of G, and $e = |E|$ = size of G.

For example, the graph above has order 5 because it consists of five vertices, and has size 6 because it has a total of six edges: AB, AC, AD, BC, CD and CE.

Definition

The number of edges incident to a vertex is called the **degree of the vertex**. The degree of a vertex A is denoted by $deg(\text{A})$.

When you want to find the degree of a vertex, you just need to count the number of edges incident with this vertex. A vertex with degree zero is an isolated vertex, i.e. a vertex that is not the endpoint of any edge and therefore not adjacent to any other vertex in the graph.

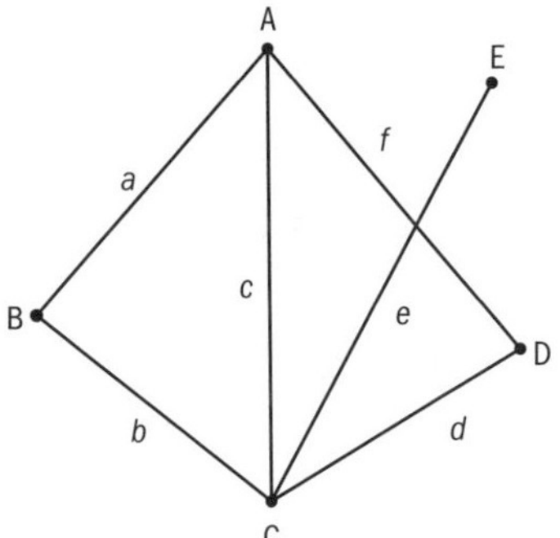

For the graph here, $deg(\text{A}) = 3$, $deg(\text{B}) = 2$, $deg(\text{C}) = 4$, $deg(\text{D}) = 2$ and $deg(\text{E}) = 1$.

Some graphs – called **multigraphs** - may include **multiedges** and **loops**, as illustrated in the diagram below. Multiedges are two or more edges connecting the same pair of distinct vertices; a loop is an edge joining a vertex to itself. Loops are counted twice towards the degree of their endpoints. In this diagram, the vertex C has degree 6.

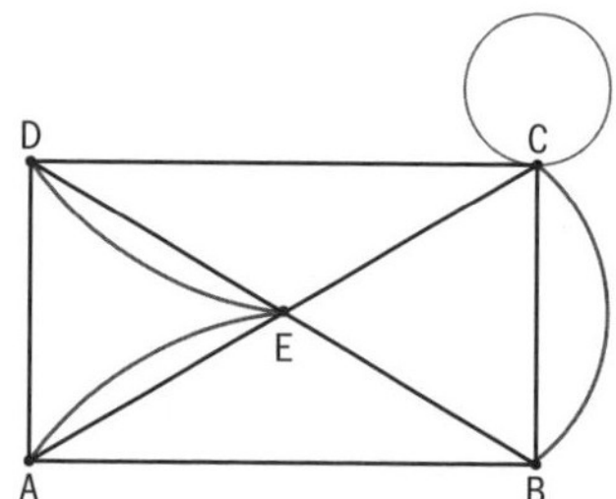

An **adjacency table** $\{a_{ij}\}$ shows whether or not two given vertices V_i and V_j in a graph are connected:

- if they are connected, the value of the entry a_{ij} is the number of edges connecting them;
- otherwise $a_{ij} = 0$.

For the multigraph above, an adjacency table (taking the vertices in order A, B, C, D and E) is shown, below right.

We can read the degree of each vertex by simply adding the entries of the corresponding row or column. For example, A has degree $0 + 1 + 0 + 1 + 2 = 4$, and C has degree $0 + 2 + 2 + 1 + 1 = 6$.

We can also find the order of the graph by counting the number of columns or rows of the table (5 in this case), and find the size of the graph by adding all the entries in the table and then dividing it by 2 since each edge is incident with two vertices (and a loop is counted twice). In this case the size is 12.

	A	B	C	D	E
A	0	1	0	1	2
B	1	0	2	0	1
C	0	2	2	1	1
D	1	0	1	0	2
E	2	1	1	2	0

Example 1

For the graph here, write down:

a the number of vertices **b** the number of edges

c the degree of each vertex **d** the adjacency table.

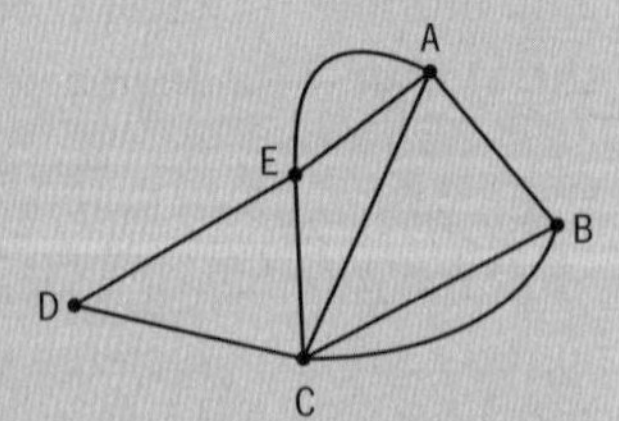

a 5 vertices

Count the points labelled A, B, C, D and E.

b 9 edges

Count the arcs connecting the vertices.

c $deg(\text{A}) = 4$, $deg(\text{B}) = 3$, $deg(\text{C}) = 5$, $deg(\text{D}) = 2$ and $deg(\text{E}) = 4$

Count the number of edges incident with each vertex.

d

	A	B	C	D	E
A	0	1	1	0	2
B	1	0	2	0	0
C	1	2	0	1	1
D	0	0	1	0	1
E	2	0	1	1	0

If two vertices in a graph are connected, the value of the entry a_{ij} is the number of edges connecting them; otherwise $a_{ij} = 0$.

Short Investigation

1 Draw several graphs and multigraphs with order 4, 5 and 6. For each of them find its size, its order and the degree of each of its vertices.

2 Investigate possible relationships between the size of the graph and the degrees of the vertices. If necessary, draw a few more graphs of different orders.

3 Investigate graphs with vertices of odd degree.

4 How many vertices of odd degree do they have? Can you draw a graph with exactly 3 vertices of odd degree? Why or why not?

This short investigation reveals a very useful theorem known as the **Handshaking lemma** that we will now prove:

> **Theorem 1**
>
> The sum of the degrees of the vertices of a graph G is twice the size of G, i.e. if $G = (V, E)$ and $|E| = e$ then $\sum_{A \in V} deg(\text{A}) = 2e$.

The name Handshaking lemma comes from the fact that we can think of the vertices as people and the handshakes as edges. Informally, we can state the Handshaking lemma as 'if *n* people meet and shake hands, the sum of individual handshakes is twice the number of the total handshakes that occurred'.

Proof:

Every edge in a graph G connects two vertices. Therefore if we add all the vertices' degrees we count each edge twice, i.e. the sum of the degrees of all the vertices of G equals twice the size of G (total number of edges). Q.E.D.

The Handshaking lemma has a very useful corollary:

> **Corollary**
>
> The number of vertices of odd degree in a graph $G = (V, E)$ is always even.

Proof:

The degree of a vertex is either odd or even. Let V_O be the set of all vertices of an odd degree and V_E be the set of all vertices of an even degree. Since $V_O \cap V_E = \varnothing$ and $V_O \cup V_E = V$, we can write

$$2e = \sum_{A \in V} deg\,(A) = \underbrace{\sum_{A \in V_O} deg\,(A)}_{m} + \underbrace{\sum_{A \in V_E} deg\,(A)}_{n} \cdot$$

As n must be even, as it is the sum of even numbers (the degrees of the vertices of even degree), m must also be even, as $m + n = 2e$. Therefore, the number of odd degrees added to obtain m must be even. Q.E.D.

> **Definition**
>
> $H = (V', E')$ is called a **subgraph** of a graph $G = (V, E)$ if and only if $V' \subseteq V$ and $E' \subseteq E$ where $V' \neq \varnothing$, and the edges in E' are adjacent to the vertices in V'.

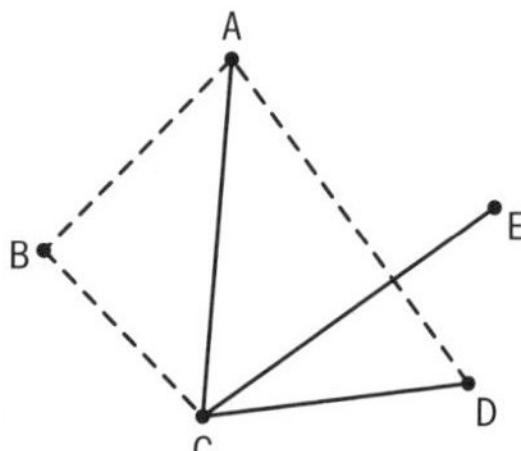

For example, the diagram to the right shows with dashed lines the graph $H = (V', E')$ as a subgraph of $G = (V, E)$ because $V' = \{A, B, C, D\} \subseteq V = \{A, B, C, D, E\}$ and $E' = \{AB, BC, AD\} \subseteq E = \{AB, AC, AD, BC, CD, CE\}$.

Exercise 4A

1 For the graph in the diagram, write down:

- **a** the number of vertices
- **b** the number of edges
- **c** the degree of each vertex
- **d** the adjacency table.

2 Draw the graphs described by the following adjacency tables. State the degree of each vertex, the size of the graph and its order.

a

	A	B	C	D
A	0	1	1	0
B	1	0	1	0
C	1	1	0	1
D	0	0	1	0

b

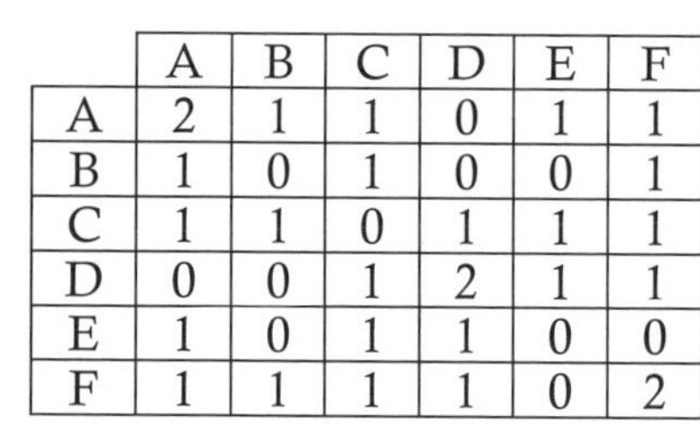

	A	B	C	D	E	F
A	2	1	1	0	1	1
B	1	0	1	0	0	1
C	1	1	0	1	1	1
D	0	0	1	2	1	1
E	1	0	1	1	0	0
F	1	1	1	1	0	2

3 The graphs $G_1 = (V_1, E_1)$ and $G_2 = (V_2, E_2)$ have the following properties: $|V_1| = |V_2| = 6$, $|E_1| = 10$ and $|E_2| = 12$.

a Draw a pair of possible graphs G_1 and G_2.

b Write down the adjacency tables for G_1 and G_2.

4 A graph contains 22 vertices and 43 edges. Every vertex has a degree of 3 or 5. Find the number of vertices with degree 3 and the number of vertices with degree 5.

Graphs are used in the field of chemistry to model chemical compounds. In computational biochemistry some sequences of cell samples must be excluded to resolve the conflicts between two sequences. This is modelled in the form of a graph where the vertices represent the sequences in the sample. An edge will be drawn between two vertices if and only if there is a conflict between the corresponding sequences. The aim is to obtain subgraphs by removing vertices to eliminate all conflicts in the sequences.

4.2 Classification of graphs

In this section we are going to study the most usual and useful classifications of graphs. Be aware that the same graph can be classified according to different properties.

Weighted graphs

If the edges of a graph are assigned a number (the weight) we say that the graph is a **weighted graph** like the ones shown here.

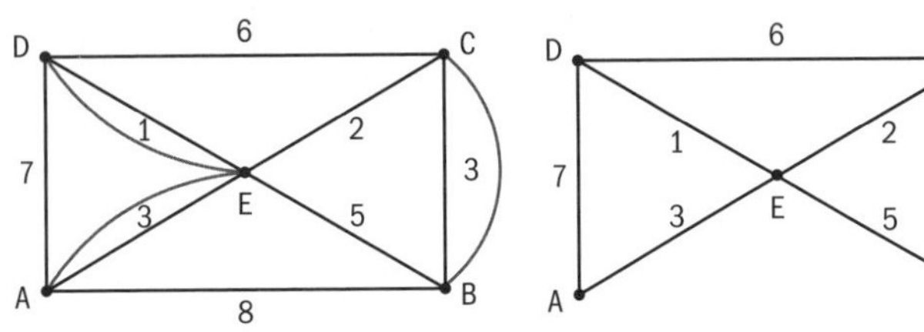

We will explore weighted graphs in detail in chapter 5. In general, the weight of a graph represents a quantity that people want to minimize, for example: the cost of a flight, length of a road or time necessary to complete a task. In chapter 5 we will look at strategies to obtain optimal solutions to such practical problems.

Directed graphs

A **directed graph** or digraph is a set of vertices connected by edges, where each edge has a direction associated with it. Digraphs can also be weighted graphs.

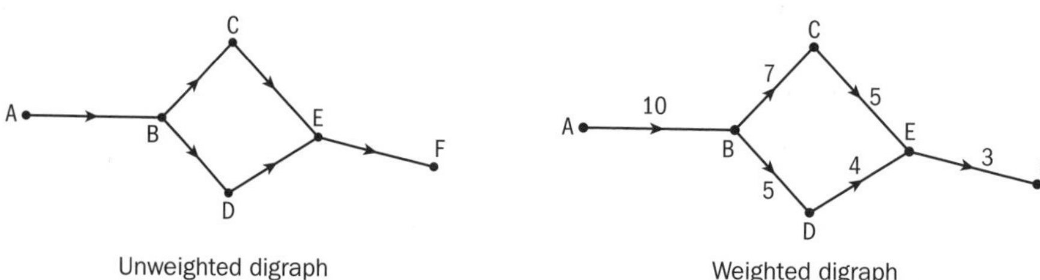

Unweighted digraph

Weighted digraph

Simple graphs

> **Definition**
>
> A **simple graph** is an unweighted, undirected graph containing no loops nor multiple edges.

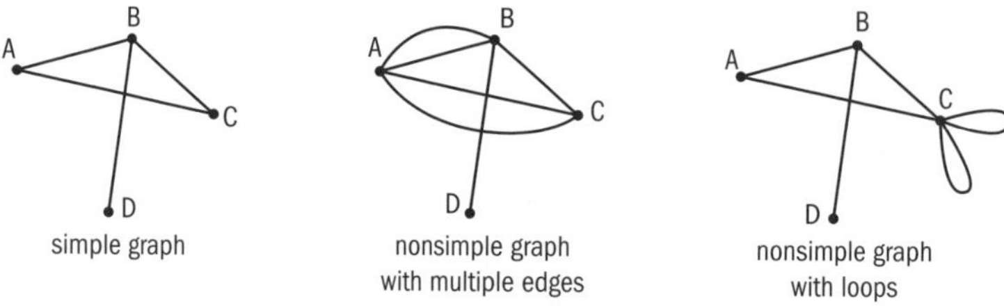

simple graph

nonsimple graph with multiple edges

nonsimple graph with loops

In order to classify simple graphs further you need to learn a few more definitions:

> **Definitions**
>
> A **walk** is a sequence of linked edges. Usually we describe a walk by listing the vertices in order as we walk it.
>
> The **length of a walk** is its total number of edges listed.
>
> A **path** is a walk with no repeated vertices.
>
> The **degree sequence** of a path lists the degrees of the vertices in the order we pass through them as we walk along the path.
>
> A **cycle** is walk that begins and ends at the same vertex, and has no repeated vertices. The **length of a cycle** is its total number of distinct vertices listed.
>
> A **trail** is a walk with no repeated edges.
>
> A **circuit** is a walk that begins and ends at the same vertex, and has no repeated edges.

The number of vertices listed to describe a walk is one more than its length.

Example 2

The graph G is shown here.

a Explain why it is a simple graph.

b Find two distinct paths connecting A and D and their corresponding degree sequence.

c State two cycles of G and their length.

d State a walk that is a circuit.

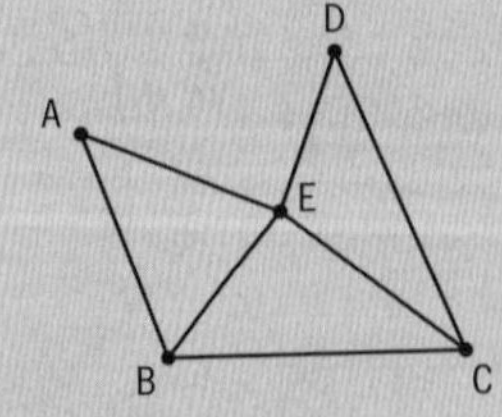

a G is an unweighted, undirected graph containing no loops nor multiple edges.	*Use the definition of a simple graph.*
b AED with degree sequence 2, 4, 2; ABED with degree sequence 2, 3, 4, 2.	*The path AED has edges AE and ED. The path ABED has edges AB, BE and ED. List in order the degrees of the vertices that describe the path.*
c ABEA is a cycle of length 3; BCDEB is a cycle of length 4.	*Select a path such that the first vertex of the path corresponds to the last, and count the number of distinct vertices.*
d ABECDEA is circuit.	*Select a walk without repeated edges such that the first vertex corresponds to the last. Note that you may walk through the same vertex more than once.*

Connected graphs

A graph is **connected** when there is a path from any vertex to any other vertex in the graph. A graph that is not connected is said to be **disconnected**. The following diagram shows all possible connected simple graphs with 2, 3 and 4 vertices respectively. The empty graph and the graph with a single vertex are considered trivial cases of connected graphs.

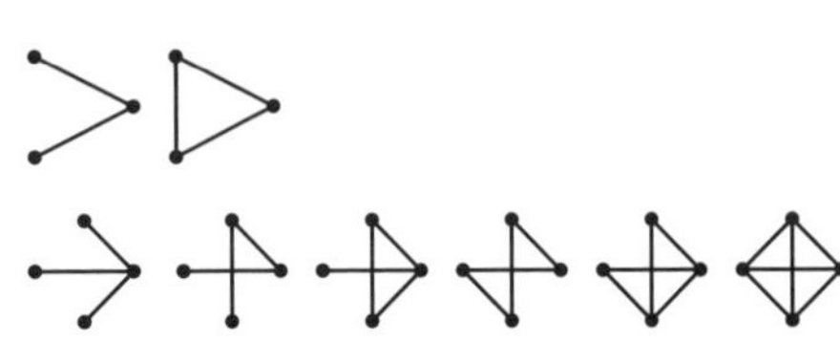

The Pigeonhole Principle, studied in chapter 2, allows us to prove the following result about simply connected graphs:

Theorem 2

Let G be a simply connected graph of order n, for $n \geq 2$. Then G has at least two vertices with the same degree.

Simply connected means that the graph is both simple and connected.

Proof:

As G is a connected graph with more than one vertex, the degree of each vertex is at least one, i.e. $deg(V_i) \geq 1$. As the graph G is simple, there are no multiple edges; therefore each vertex can be connected with at most $n - 1$ vertices, i.e. $deg(V_i) \leq n - 1$. This means that the list of possible degrees for each vertex of G is 1, 2, ..., $n - 1$. As there are n vertices, by the Pigeonhole Principle, at least two vertices must have the same degree. Q.E.D.

In Graph Theory a collection of disjoint trees is called a **forest**.

In Computer Science, trees are very useful tools, particularly rooted trees that have one vertex – the root singled-out as the starting point for all branches of the tree. Rooted trees can be used to store data in computers in different ways that require more or less memory to encode this data.

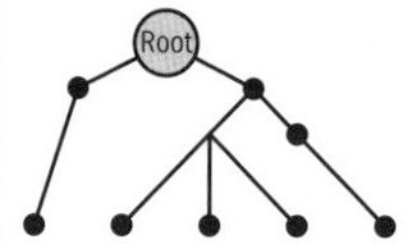

Trees

A **tree** is a connected graph with no cycles.

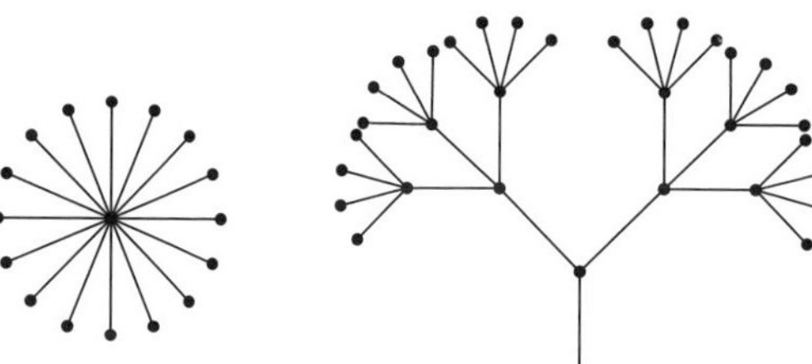

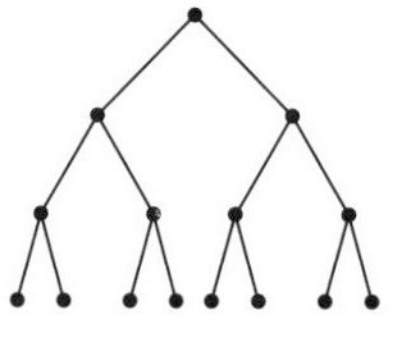

The diagram above shows a **forest** that consists of three rooted trees.

Short Investigation

Draw a tree with 5 vertices.

For any two vertices in the tree, how many paths can you find that have them as endpoints?

How many edges does the tree have?

Can you remove a vertex without disconnecting the graph?

Draw more trees with 6, 7 and 8 vertices respectively and answer the same questions.

State your conclusions.

This investigation reveals important properties of trees. In fact these properties can be used as alternative definitions of a tree as established in the following theorem:

Theorem 3

Let $G = (V, E)$ be a finite graph with more than one vertex.
The following statements are equivalent:

1. G is a tree.
2. Each pair of vertices of G is connected by exactly one path.
3. If a is an edge of G then $G - \{a\}$ is disconnected.
4. G is cycle-free and has $n - 1$ edges.
5. G is connected and has $n - 1$ edges.

Proof:

We start with the definition of tree (a connected graph with no cycles) and need to show that 1. ⇒ 2. ⇒ 3. ⇒ 4. ⇒ 5. ⇒ 1.

Consider $G = (V, E)$ with $V = \{V_1, ..., V_i, V_j, ..., V_n\}$, $n \geq 2$.

1. ⇒ 2.

G is a tree. Therefore any two vertices V_i and V_j are connected by at least one path. Suppose that two distinct paths P and Q connecting V_i and V_j exist. Then G would contain a cycle $P \cup Q$ which contradicts the assumption that G is a tree. Therefore each pair of vertices of G is connected by exactly one path.   Q.E.D.

2. ⇒ 3.

Suppose now that each pair of vertices of G is connected by exactly one path. Consider the vertices V_i and V_j connected by the edge a. $\{a\}$ is a path and is the only one connecting V_i and V_j. Therefore $G - \{a\}$ is disconnected as it contains no path connecting V_i and V_j.   Q.E.D.

The remaining proofs are left to you as an exercise.

Complete graphs

Complete graphs are a very important family of graphs and we have special notation to represent each family member:

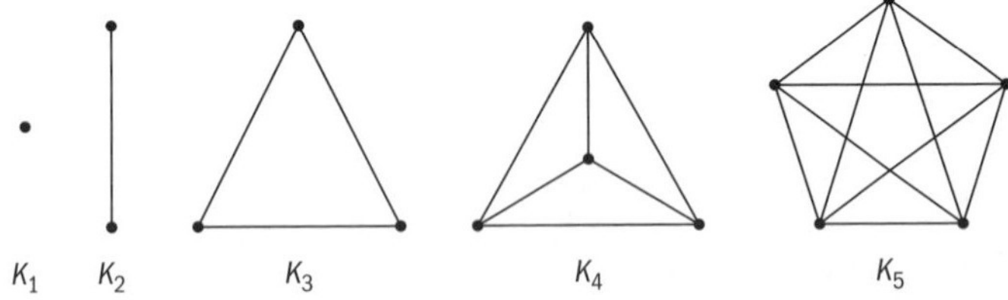

As illustrated above, the index of the letter K gives the number of vertices or order of the graph K_n. **Complete graphs** are characterized by the property: 'each vertex is adjacent to every other vertex in the graph'. This means that it is not possible to add edges to these graphs without obtaining a multigraph!

Theorem 4

Let K_n be the complete graph of order n, $n \in \mathbb{Z}^+$. The number of edges (the size) of K_n is $\frac{n(n-1)}{2}$.

Proof:

The number of vertices is n. As each edge connects two vertices, there are exactly $\binom{n}{2} = \frac{n(n-1)}{2}$ edges.   Q.E.D.

An alternative way of obtaining the size of K_n is by looking at its adjacency table.

The adjacency tables of complete graphs have 1 for all entries, except on the leading diagonal where all the entries are 0. For example the adjacency table of K_4 is:

	A	B	C	D
A	0	1	1	1
B	1	0	1	1
C	1	1	0	1
D	1	1	1	0

Therefore, the number of edges of K_n is given by $\dfrac{\overbrace{n}^{\text{number of rows}} \times \overbrace{(n-1)\times 1}^{\text{number of 1s per row}}}{2}$.

Adjacency tables of complete graphs are symmetric with respect to the main diagonal, which means that we can deduce this result using columns instead of rows.

Bipartite graphs

A simple graph $G = (V, E)$ is **bipartite** if its set of vertices V can be partitioned into two disjoint sets M and N such that each edge of G connects a vertex of M with a vertex of N, i.e. M and N are such that $M \cup N = V$ and $M \cap N = \varnothing$, often called a **partition**, and all the edges for the set E are of the form XY such that X $\in M$, Y $\in N$.

Example 3

Consider the graph G in the diagram.

a Write down its adjacency table.

b Explain how it shows that G is bipartite. State clearly the disjoint sets of vertices M and N.

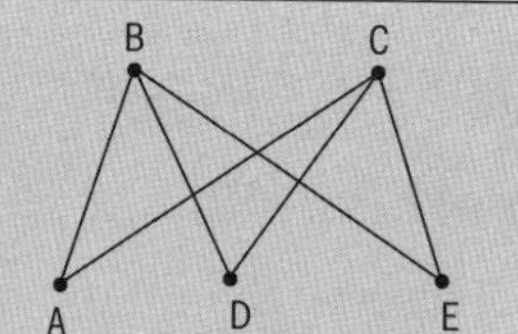

a

	A	B	C	D	E
A	0	1	1	0	0
B	1	0	0	1	1
C	1	0	0	1	1
D	0	1	1	0	0
E	0	1	1	0	0

The entry $a_{ij} = 1$ if two vertices are connected; otherwise $a_{ij} = 0$.

b $M = \{B, C\}$ and $N = \{A, D, E\}$ are the two disjoint sets because there are no connections among the vertices of these sets, as shown in the adjacency table.

Rearrange rows and columns of the adjacency table to show the split in the sets $M = \{B, C\}$ and $N = \{A, D, E\}$.

	B	C	A	D	E
B	0	0	1	1	1
C	0	0	1	1	1
A	1	1	0	0	0
D	1	1	0	0	0
E	1	1	0	0	0

Bipartite graphs are also called **2-colourable** in the sense that we could colour the vertices of each split set in a different colour. This is a useful practical method to identify bipartite graphs: if we can colour the vertices using exactly two colours so that no two vertices with same colour connect, the graph is bipartite.

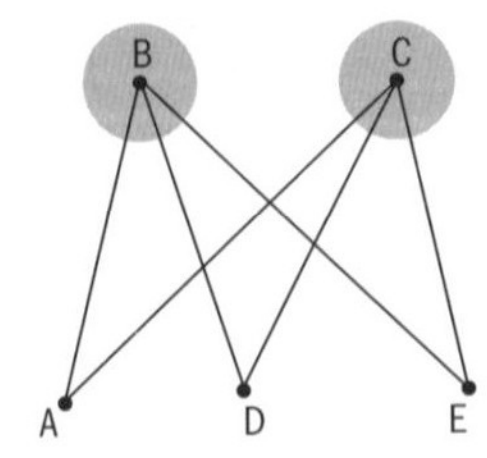

In Example 3, it is easy to show that the graph is bipartite by colouring or shading B and C.

A bipartite graph is said to be a **complete bipartite graph** if every vertex from M is adjacent to every vertex from N. The notation for a complete bipartite graph is $K_{m,n}$ where $|M| = m$ and $|N| = n$.

The diagram in Example 3 (redrawn here at the right, with shaded vertices) shows a complete bipartite graph $K_{2,3}$ in the sense that all the vertices in the set M are connected to all the vertices in the set N.

Theorem 5

The number of edges of the complete bipartite graph $K_{m,n}$ is mn, i.e. $|K_{m,n}| = mn$.

Proof:
As each edge connects a vertex from M to a vertex from N, and each vertex of M is connected to each vertex of N, the number of edges is given by $|M| \times |N| = mn$. Q.E.D.

Exercise 4B

1 Observe the following graphs:

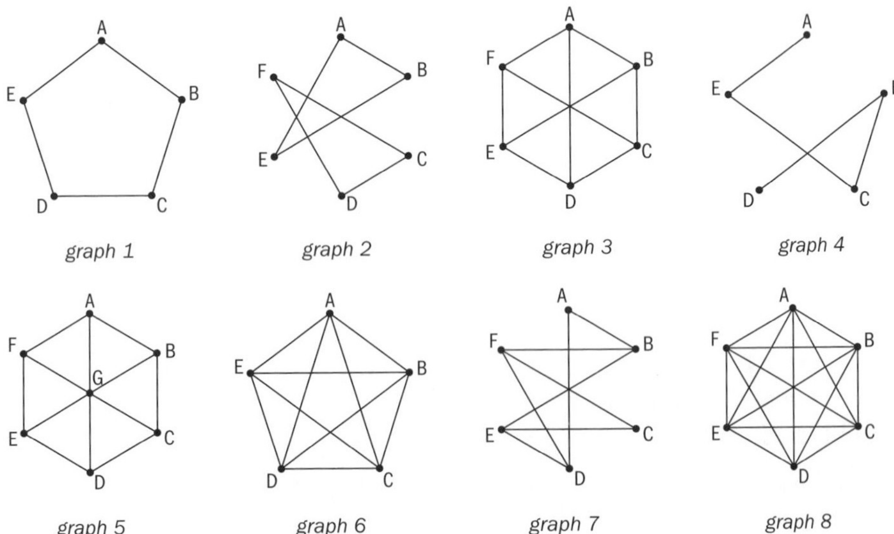

State which graphs are:

a bipartite **b** complete

c trees **d** disconnected.

2 State the order $|V|$ and size $|E|$ of each graph $G = (V, E)$ in question **1**.

3 Show that trees are bipartite graphs.

4 Prove that the following statements are equivalent:

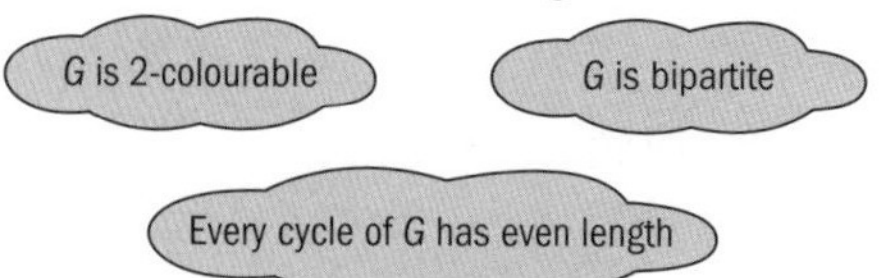

5 A simply connected graph of order 9 has all vertices with the same degree d. Find all possible values of d.

6 Find the number of vertices and edges for the following graphs.

a $K_{3,4}$ **b** $K_{13,17}$ **c** $K_{12,5}$

7 A complete bipartite graph $K_{m,n}$ has altogether 24 vertices and 128 edges. Find the number of vertices in each set of the partition.

8 Complete the proof of Theorem 3.

EXAM-STYLE QUESTION

9 Use the principle of mathematical induction to show that a tree with n vertices has exactly $n - 1$ edges.

For question 9 you may use weak or strong induction.

4.3 Different representations of the same graph

In the following sections we will need to identify properties of graphs that may require that we redraw them. When doing so we need to be careful to produce representations of the graphs in such a way that all the incidence relations are preserved. In other words, edges incident with vertices remain incident with them; vertices adjacent to other vertices remain adjacent with them.

The diagrams below show you two representations of a cube.

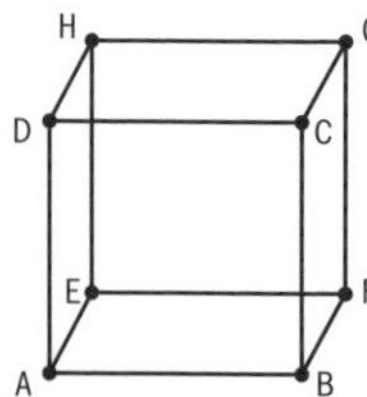

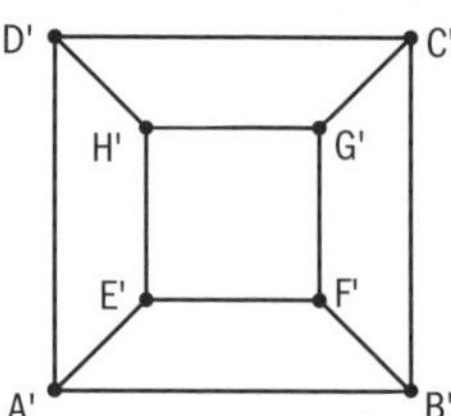

These graphs have 8 vertices, 12 edges and each of their vertices has degree 3.

When we produce two diagrams showing a solid we say that we actually draw two isomorphic graphs, i.e. graphs that have exactly the same properties like order, size, number of connected components and sequence of degrees. These properties are called isomorphism invariants.

As a formal study of isomorphic graphs is not part of the course, we will treat this topic informally.

Useful isomorphism invariants:

If two graphs G and H are isomorphic, then:

- The size of G equals the size of H.
- The order of G equals the order of H.
- The sequence of the degrees of the vertices of G is a permutation of the sequence of the degrees of the vertices of H.
- The number of connected components of G and H is equal.
- The lengths of the cycles of G matches the lengths of the cycles of H.

You need to be careful when using isomorphism invariants; even if you check a few of them you may fail to identify differences between graphs. For example, the diagram below shows two connected graphs with 8 vertices, 12 edges, and all vertices have degree 3, but they have cycles with different lengths which makes it impossible for them to be isomorphic. This means that these two graphs are distinct in Graph Theory.

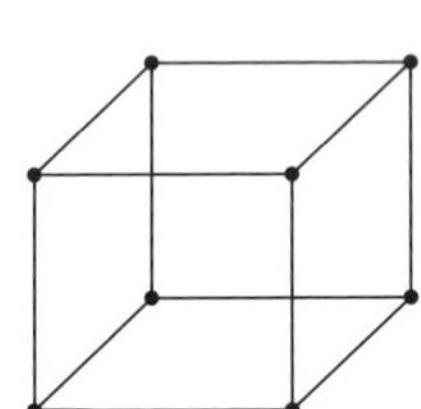

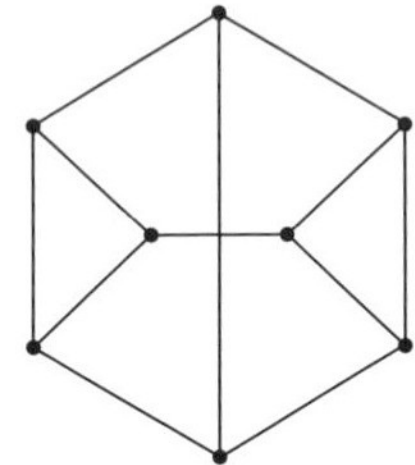

Example 4

Justify why each pair of graphs G and H cannot be isomorphic.

a

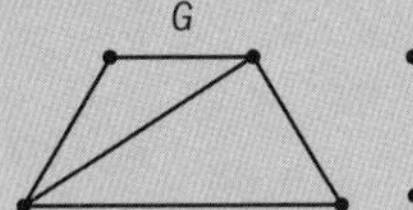

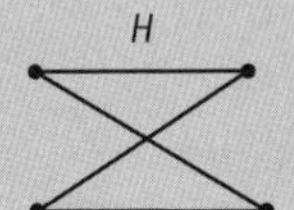

b

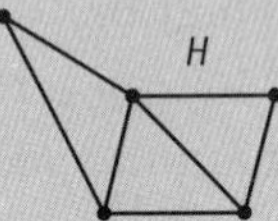

a For example, G and H have different sizes; G has two vertices with degree 3 and H has all vertices with degree 2; G has two cycles with length 3 (triangles) and H has just one cycle with length 4.	*Use isomorphism invariants to identify differences between the graphs.*
b G has four vertices with order 3 and one vertex with order 2 while H has two vertices with order 2, two vertices with order 3 and one vertex with order 4; G has a cycle with length 4 but not H.	

Exercise 4C

1 Consider the following pairs of graphs. In each case, decide whether or not each pair are isomorphic. In the case that they are isomorphic, define an isomorphism f between them; otherwise provide a justification as to why such an isomorphism cannot exist.

a

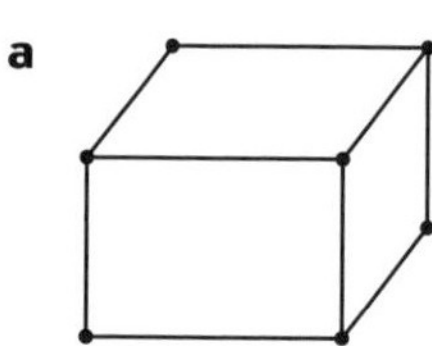

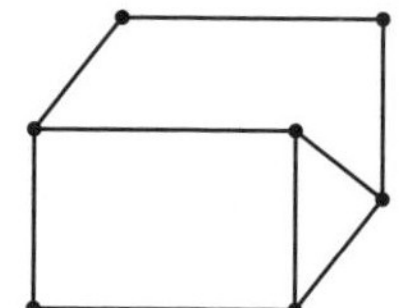

b

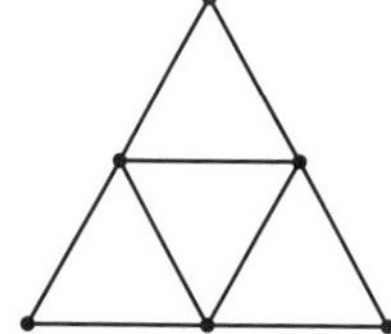

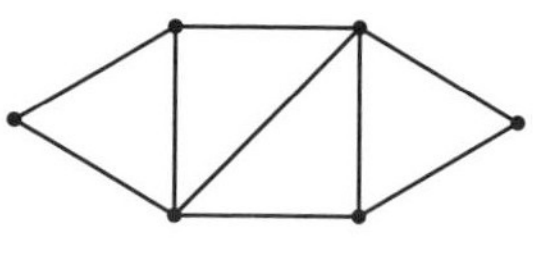

c

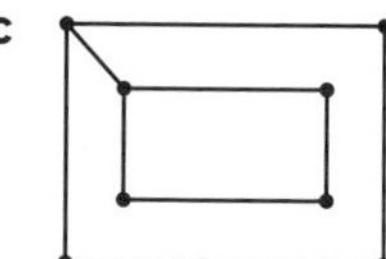

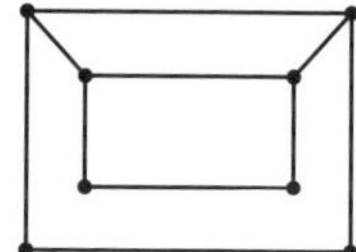

2 Draw two non-isomorphic graphs with 3 vertices and 2 edges. How many such non-isomorphic graphs are possible?

3 Draw two non-isomorphic graphs with 4 vertices and 3 edges. How many such non-isomorphic graphs are possible?

4 If all the vertices of a graph have the same degree we say the graph is **regular**.
Draw all possible non-isomorphic simple regular graphs with 4 vertices.

5 A graph is said to be **coloured** with n colours if a colour can be assigned to each vertex in such a way that every vertex has a colour which is different from the colours of all its adjacent vertices. Show that the complete graph K_n requires n colours to be coloured.

4.4 Planar graphs

Definition
A graph G is **planar** if and only if we can draw it in a plane without any edges crossing over each other.

Such drawing of a planar graph G is a graph *H* isomorphic to G called a **planar embedding** of G.

Cycles of planar graphs divide the plane into regions called **faces.** This name comes from the planar representations of polyhedra like the cube below. As you know, a cube has 6 faces and each face is a square. The planar representation of a cube shows 6 cycles of length 4, each of them enclosing a face. Note that one of these faces is the region outside the graph.

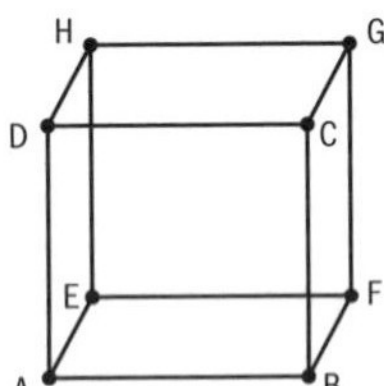

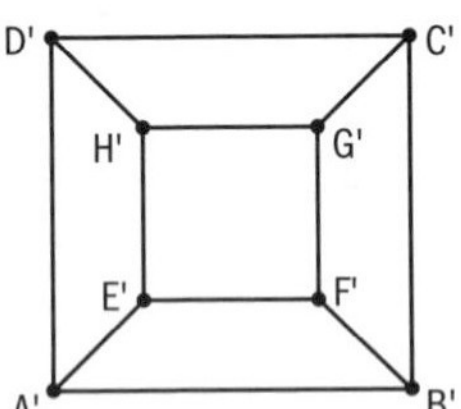

All polyhedra can be represented by planar graphs. For example, the five Platonic solids can be represented by the graphs below:

tetrahedron

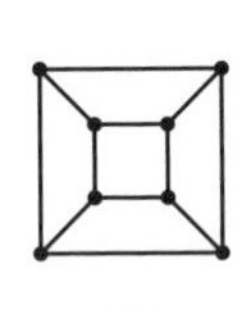
cube

octahedron

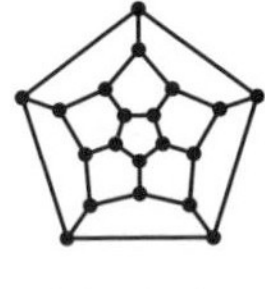
dodecahedron

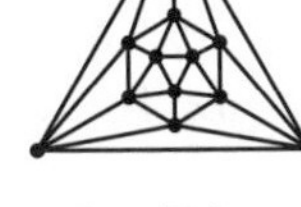
icosahedron

All these graphs are called **regular** as all the vertices have the same degree.

The Ancient Greek Plato described how he thought the five regular solids, now called the Platonic Solids, make up the four elements (fire, earth, water and air) and heaven. However, Platonic solids were used as art motifs even before Plato and they remained popular during the Renaissance among mathematicians and artists like Piero della Francesca, Luca Pacioli, and Leonardo da Vinci.

Example 5

Show that K_1, K_2, K_3, K_4 are planar.

K_1 K_2 K_3 K_4

Draw planar graphs of each graph.

We will show that K_n for $n \geq 5$ is not planar. Before we proceed with this proof we need to introduce a few more concepts and results.

Spanning trees

Definition

A subgraph T of a graph G is called a **spanning tree** if T is a tree and includes all the vertices of G.

Example 6

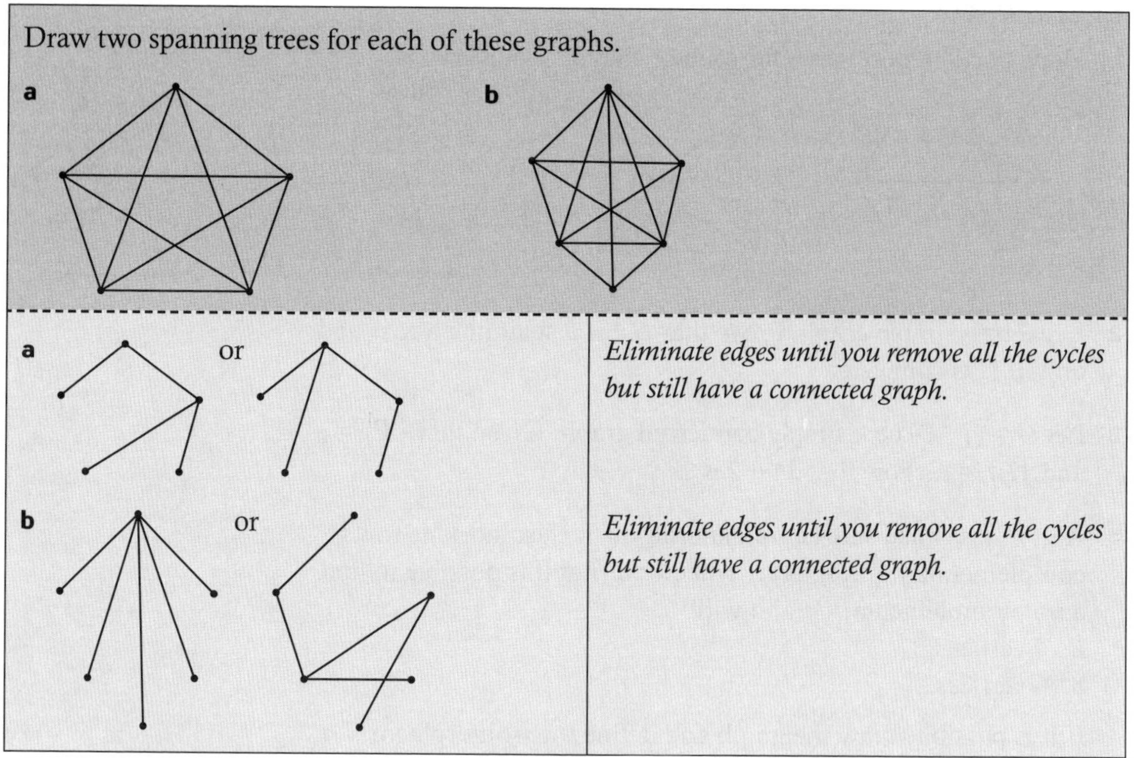

Complements of graphs

Let $G = (V, E)$ be a simple graph of order n. Consider the representation of K_n that has the vertices of G. Then the **complement** of G, denoted by G', is a graph that contains the same set of n vertices V as the graph G, and contains all the edges of K_n that G does not contain.

The diagram below shows you a graph G with order 5, K_5, and G' (the complement of G).

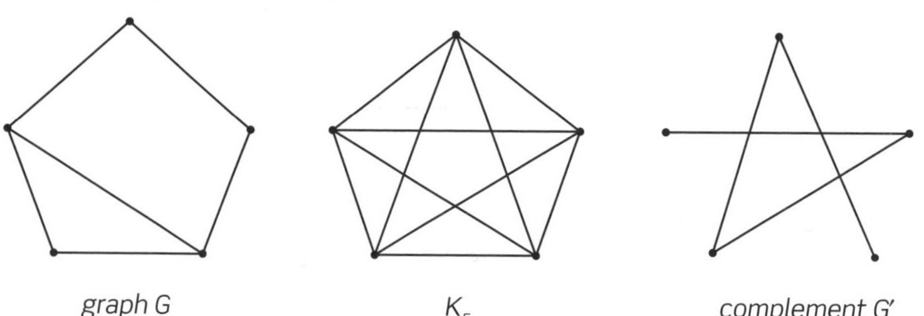

Note the number of edges of G together with the number of edges of G' is equal to the number of edges of K_n. This means that the complement of K_n consists of all the vertices and no edges, i.e. a **null graph** or **edgeless graph**.

In Graph Theory the term *null graph* may appear with different meanings: sometimes null graph means order zero graph, i.e. the graph K_0 with no edges and no vertices. In the context of the complement of a graph of order n, however, null graph means simply no edges, as the number of vertices must be n.

Exercise 4D

1 Draw two spanning trees for each of the following graphs:

a

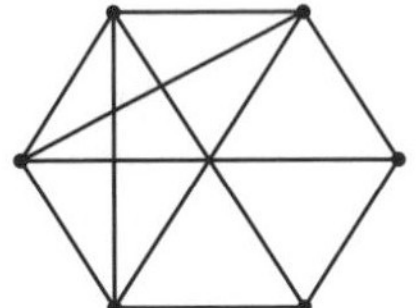

b

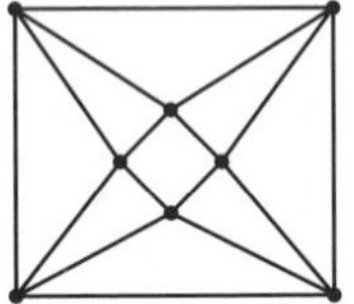

c

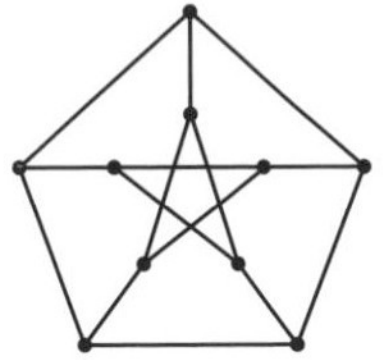

d

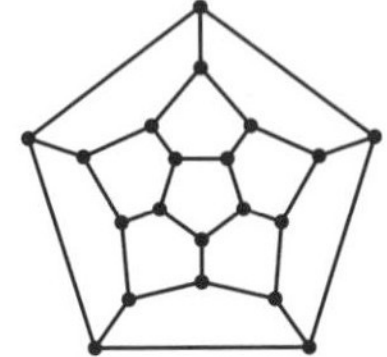

2 Let G be a simple graph. Prove that G has a spanning tree if and only if G is connected.

3 Let $G = (V, E)$ be a simple connected graph. Given that $|V| = v$ and $|E| = e$, show that $2v - 2 \leq 2e \leq v^2 - v$.

4 A graph is called **self-complementary** if it is isomorphic to its complementary graph. State whether or not it is possible to find a self-complementary graph with:

a 4 vertices

b 6 vertices.

If it is possible, draw the graph and define the isomorphism between the graph and its complement.

Euler relation for planar graphs

Mini Investigation

Using models of polyhedra, i.e. solids with flat faces, straight edges and sharp corners (for example objects with the shape of cuboids, pyramids or prisms), create a table with the following headings:

Polyhedron	Number of faces (f)	Number of edges (e)	Number of vertices (v)	$f + v$
Cube	6	12	8	14

Complete the table using as many different polyhedra as you can find. Then compare the values in the 3rd and 5th columns and write down a conjecture.

The result obtained in the mini investigation is known as the Euler formula for polyhedra or planar graphs, because all polyhedra can be represented by a planar graph: one of its nets. The Euler formula provides a relation between the number of faces, edges and vertices of any graph that can be drawn on the plane without crossing edges.

The expression $v - e + f$ in Theorem 6 is called the Euler characteristic for polyhedra.

Theorem 6

Given a connected planar graph G, the number of vertices v, edges e and faces f satisfy the formula: $v - e + f = 2$.

Proof:

Consider a spanning tree T of G. A tree has exactly one face: $f = 1$.
A tree with n vertices has $n - 1$ edges, thus $e = v - 1$.
So, $v - e + f = v - (v - 1) + 1 = 2$.

To obtain G from T we need to add edges. For each edge added, the number of faces increases by 1, leaving the Euler characteristic unchanged. Q.E.D.

Corollary 1

Let G be a connected planar graph with at least 3 vertices.
Then $e \leq 3v - 6$.

Proof:

If G is a tree with at least 3 vertices, then the result follows easily. Note that the result is false if $v < 3$.

If G is not a tree then it contains at least one cycle, which means that each face is bounded by at least 3 edges.

So, $3f \leq 2e$.

The result follows now from Euler's Theorem:

$$v - e + f = 2 \Rightarrow 3v - 3e + 3f = 6$$

$$3f \leq 2e \Rightarrow 3v - 3e + 2e \geq 6$$

$$\therefore e \leq 3v - 6 \quad \text{Q.E.D.}$$

? The work of Swiss mathematician Leonhard Euler covers so many fields in mathematics and he is often the earliest written reference on a given matter. For this reason there are many theorems and formulas associated to his name. The Euler characteristic for polyhedra is an important example. This formula allows us to prove the existence of exactly 5 Platonic solids.

Corollary 2

Let G be a connected planar graph with no triangles. Then $e \leq 2v - 4$.

A planar graph with no triangles is a planar graph with no cycles of length 3.

Proof:

We proceed as before but this time the minimum length of a cycle is 4. So, each face is bounded by at least 4 edges.

So, $4f \leq 2e$.

The result follows now from Euler's Theorem:

$v - e + f = 2 \Rightarrow 4v - 4e + 4f = 8 \Rightarrow 4v - 4e + 2e \geq 8$

$\therefore\ e \leq 2v - 4$ Q.E.D.

These two corollaries of Theorem 6 allow us to prove two important theorems.

Theorem 7

K_5 is not planar.

Proof:

K_5 has 5 vertices and 10 edges. By Corollary 1, $e \leq 3v - 6$. However, $10 \leq 3 \times 5 - 6 \Rightarrow 10 \leq 9$, which is a false statement. Therefore K_5 cannot be planar.

Theorem 8

$K_{3,3}$ is not planar.

Proof:

$K_{3,3}$ does not contain a triangle since it is a bipartite graph and all its cycles have length of at least 4. By Corollary 2, $e \leq 2v - 4$. The number of vertices is 6 while the number of edges is 9, therefore $9 \leq 2 \times 6 - 4 \Rightarrow 9 \leq 8$, which is a false statement. Therefore $K_{3,3}$ cannot be planar.   Q.E.D.

If we remove an edge from a graph, let's call it AB, and we add another vertex C together with the edges AC and BC, such an operation is called an **elementary subdivision** as, graphically, they correspond to placing a new vertex on the edge AB. Graphs are called **homeomorphic** if they can be obtained from the same graph by a sequence of elementary subdivisions.

The following theorem constitutes a very useful **criterion for planarity** and was established by Kazimierz Kuratowski in 1930.

The proof of this theorem is not included here, as its scope goes beyond what is required by the syllabus.

Theorem 9 (Kuratowski's theorem):

A finite graph is planar if and only if it has no subgraph homeomorphic to K_5 or $K_{3,3}$.

Corollary

K_n is not planar for $n \geq 5$.

Example 7

Show that the graph to the right (a Petersen graph) has a subgraph homeomorphic to $K_{3,3}$.

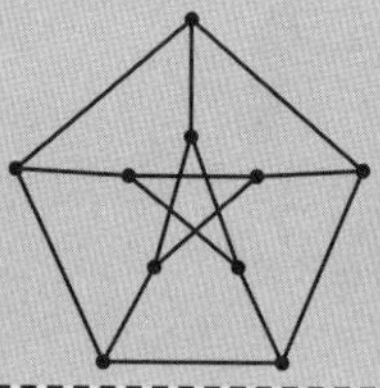

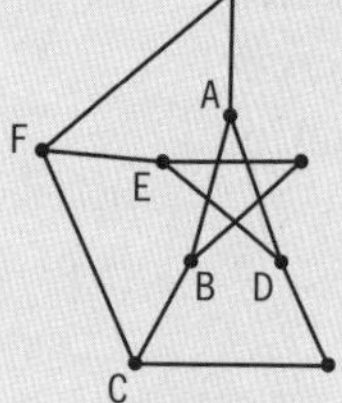

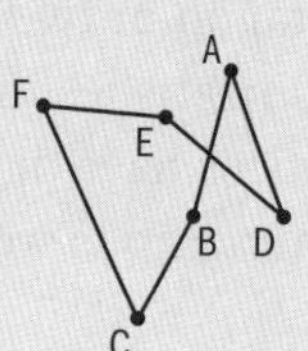

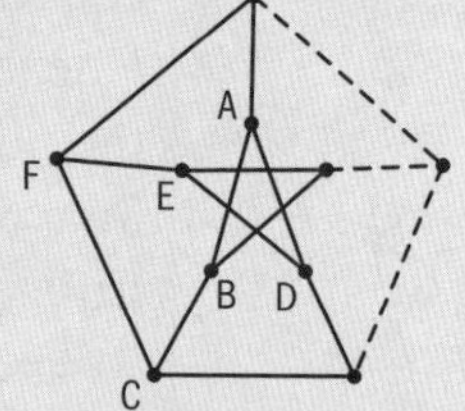

Remove the dashed edges and vertices that are not labelled.

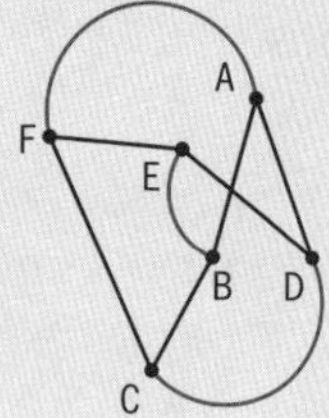

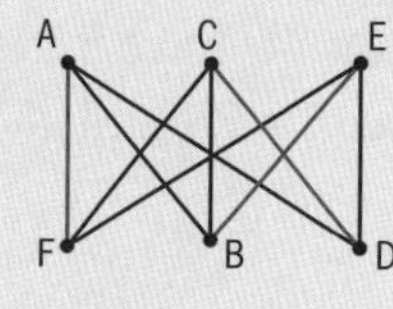

Consider the subgraph of the Petersen graph with vertices A, B, C, D, E and F which is isomorphic to $K_{3,3}$.
Note that if you start from $K_{3,3}$ and reverse the process you can obtain the Petersen graph through elementary subdivision.

There are non-planar graphs but all the graphs are spatial in the sense that they can be embedded in 3D space without having edges crossing. In 3D, the planar graphs are exactly the ones that can be represented by polyhedral solids (but not necessarily regular polyhedra) or, equivalently, the ones that can be drawn on the surface of a sphere without having crossing edges. Regular and semi-regular polyhedral solids have beautiful planar representations due to their symmetry.

Real life application – The soccer ball

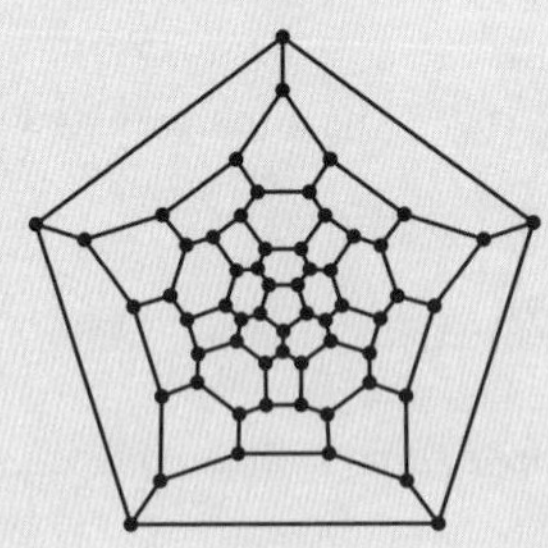

The soccer ball has evolved much in the past 50 years. The original soccer ball was heavy and made of leather strips that absorbed water and made it difficult for headers. The heavy leather laces were also painful during headers. The 1970 World Cup introduced the modern ball, known as the "Buckyball", made by the American architect and designer Richard Buckminster Fuller. Fuller's ball, the classic black-and-white, was covered with regular hexagons and pentagons rather than the old leather strips.

From a mathematical point of view, the Buckyball is a very interesting object. As the surface of a ball is covered by polygons, it can clearly be represented by a planar graph. Let p be the number of regular pentagons and h be the number of regular hexagons covering the ball. Then the number of faces on the graph is $p + h$; the number of edges e satisfies the equation $2e = 5p + 6h$ as each edge separates a hexagon from a pentagon or two hexagons; the number of vertices v satisfies the equation $3v = 5p + 6h$ as all vertices have degree 3. Using the Euler formula for planar graphs we get $v - e + f = 2$

$$\Rightarrow 6v - 6e + 6f = 12$$

$$\Rightarrow (10p + 12h) - (15p + 18h) + 6(p + h) = 12$$

$$\Rightarrow p = 12$$

Around each pentagon we have 5 hexagons, so $5p = 3h$. Therefore, $h = 20$ and

$$2e = 5 \times 12 + 6 \times 20 \Rightarrow e = 90$$

$$3v = 5 \times 12 + 6 \times 20 \Rightarrow v = 60.$$

So the Buckyball has 60 vertices, 90 edges and 32 faces (of which 12 are regular pentagons and 20 are regular hexagons). Take a classic black and white soccer ball and confirm these values. You may want to look at the new Brazuca ball too. Is it an interesting mathematical object?

Exercise 4E

1 The diagrams show two planar graphs G and H.

a State the number of vertices and the number of edges for each.

b Use Euler's formula to determine the number of faces of each graph.

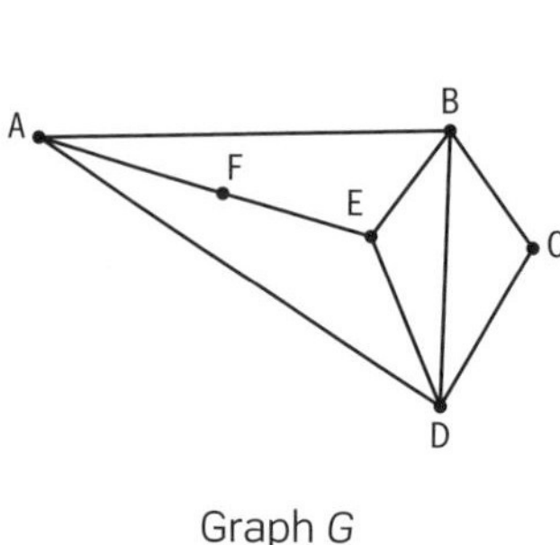

Graph *G*

Graph *H*

2 a Show that graph G in the diagram is planar.

> Use dynamic geometric software to draw a copy of the graph. Move the vertices around until no edges cross. The result is an isomorphic graph.

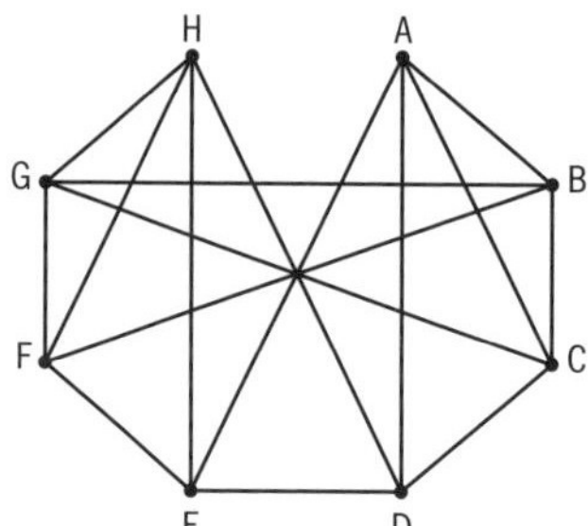

b Draw a planar representation H of G and state clearly the isomorphism that maps G onto H.

3 Consider the following graph G.

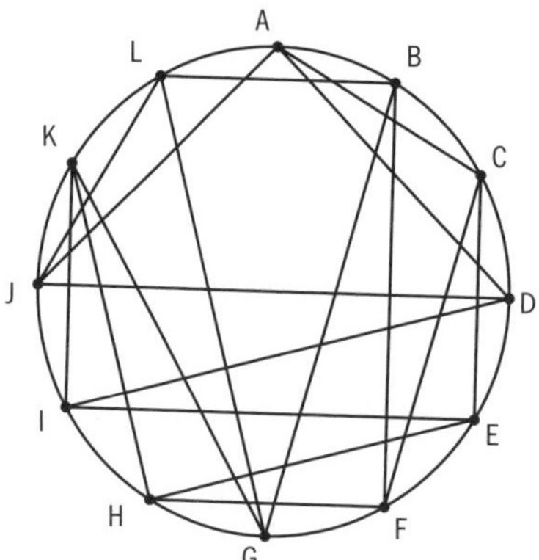

a Show that G has a planar representation.

b Hence state the size of the graph and its order.

c Use Euler's formula to find the number of faces.

d State the name of the regular polyhedron that can have the graph G as its representation. Justify your answer.

4 Consider the graphs drawn below:

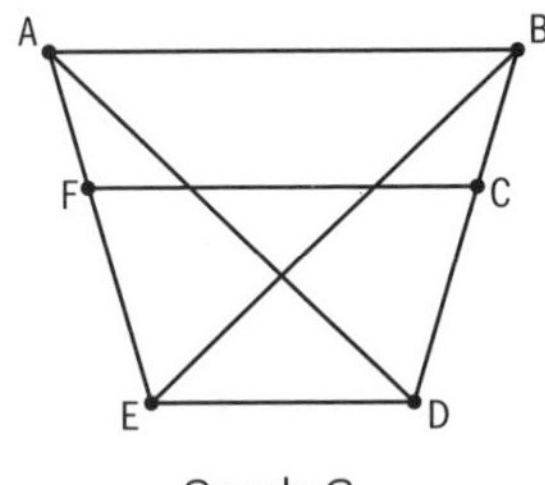

Graph G_1

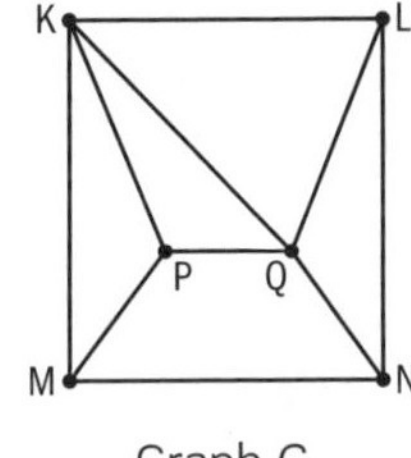

Graph G_2

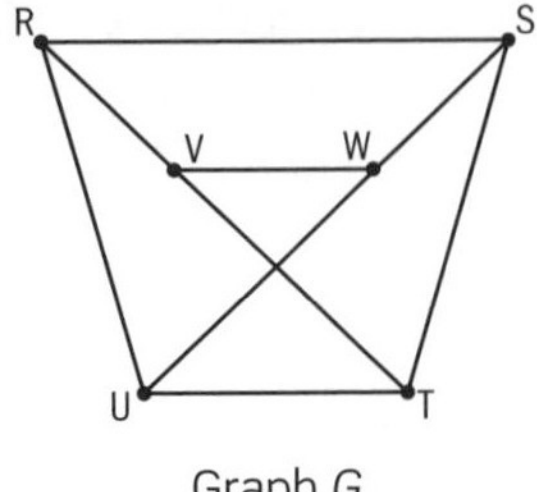

Graph G_3

a Giving a reason, determine whether or not:

i G_1 is bipartite

ii G_1 is isomorphic to G_2

iii G_3 is planar.

b Draw the complement of G_1 and state its size.

c Explain why graph G_2 can be represented by a polyhedron. Describe its faces. (You may want to sketch it first.)

5 A dodecahedron is a regular solid with 12 pentagonal faces.

a State the number of edges of a dodecahedron.

b Use Euler's formula to determine the number of vertices of this solid.

c Draw the planar representation of a dodecahedron.

EXAM STYLE QUESTION

6 **a** A simple graph G has e edges and v vertices, where $v > 2$. Prove that if all the vertices have degree at least k, then $2e \geq kv$.

b Hence prove that every planar graph has at least one vertex of degree less than 6.

4.5 Hamiltonian cycles

In 1859, the Irish mathematician Sir William Rowan Hamilton devised a puzzle with a regular dodecahedron — one of the Platonic solids studied in section 4.4. Each of the 20 vertices was labelled with a different city of the world. The goal was to start at a city and travel along the edges of the dodecahedron, visiting each of the other 19 cities exactly once, and return back at the first city. Hamilton's game was made of wood and the journey was marked off using a string and pegs.

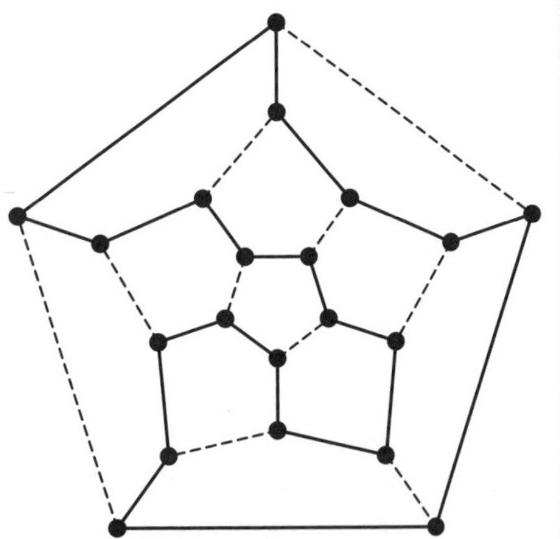

In this section we will be looking at a way of moving around a graph and introducing an important problem that will be extended further in chapter 5: can we always move around a graph and return to the starting point without visiting any vertex twice? The answer is clearly no. For example, if we have a tree we cannot return to the starting point without retracing the same route and revisiting the same vertices (and edges) again.

In order to discuss this problem you may need to recall the definitions of walk, path and cycle because precise language is a particularly important requirement in this area of the course.

Clearly we need to reformulate the question, and ask for the conditions required for such a walk around the graph to result in a special cycle: a Hamiltonian cycle, with no repeated vertices.

So, when is it possible to find a Hamiltonian path or cycle? If it exists, is it unique? Is there a method to find a Hamiltonian cycle systematically?

We know for sure that K_n for $n \geq 3$ is Hamiltonian in the sense that it is possible to walk around the graph and visit each vertex exactly once and return to the starting point. In fact, as all vertices are connected we have $n!$ different possible Hamiltonian cycles if the order in which the vertices are visited is considered. For example, for K_4 in the diagram, we can visit the following Hamiltonian cycles:

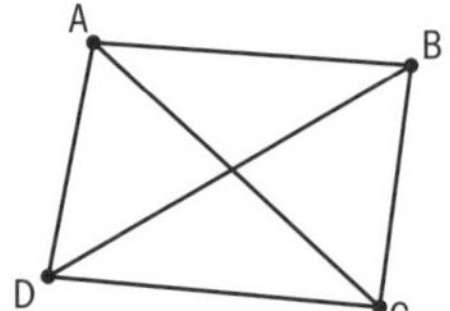

ABCDA, ABDCA, ACBDA, ACDBA, ADBCA, ADCBA
BACDB, BADCB, BCADB, BCDAB, BDACB, BDCAB,
CABDC, CADBC, CBADC, CBDAC, CDABC, CDBAC,
DABCD, DACBD, DBACD, DBCAD, DCABD and DCBAD.

However, if we delete two edges the graph will no longer have Hamiltonian cycles. For example:

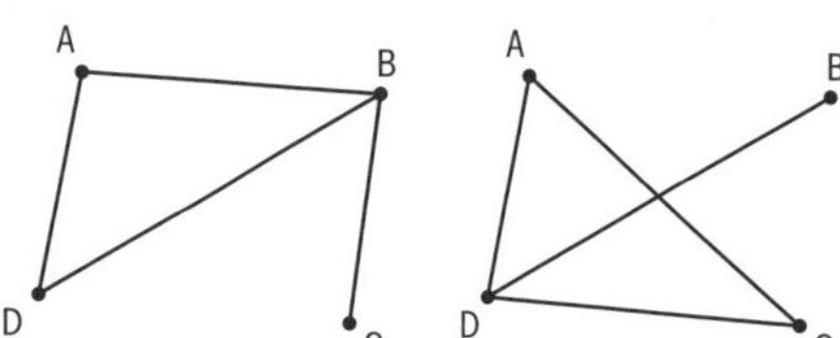

Dirac's theorem, whose proof goes beyond of the requirements of the syllabus, provides a sufficient condition on the number of edges incident with each vertex of the graph for it to have Hamiltonian cycles:

Theorem 10 (Dirac's theorem):

Let $G = (V, E)$ be a simple connected graph. If $|V| = n, n \geq 3$ and for each vertex $V_i \in V$, $\deg(V_i) \geq \frac{n}{2}$ then the graph G has a Hamiltonian cycle.

Look at the two previous diagrams; in both cases one of the vertices has degree 1 which is less than half of the order of the graph.

However, this fact alone does not allow us to conclude that the graphs are not Hamiltonian as Dirac's theorem provides only a **sufficient** condition, but not **necessary** conditions, for the existence of Hamiltonian cycles. Also, once we establish the existence of a Hamiltonian cycle there is no systematic way, i.e. an algorithm that can be executed by a computer, for finding Hamiltonian paths or cycles. Let's look at a few examples of graphs and try to decide whether or not they are Hamiltonian and, if so, find these paths and determine the number of possible cycles.

Gabriel Andrew Dirac (1925–1984) was born in Hungary and later adopted by physicist Paul Dirac, Lucasian professor of mathematics at the University of Cambridge and Nobel prize winner for Physics in 1933. Gabriel Dirac was a pioneer in the development of Graph Theory. He became a professor of pure mathematics at the University of Aarhus in Denmark. Gabriel Dirac was a dedicated teacher and a source of inspiration for other mathematicians and researchers.

Example 8

Show that these two graphs have Hamiltonian paths. State which graphs have also Hamiltonian cycles.

1

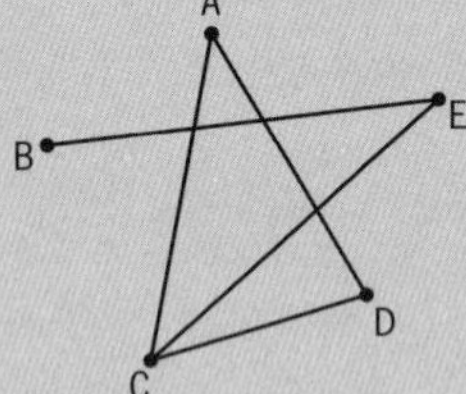

2

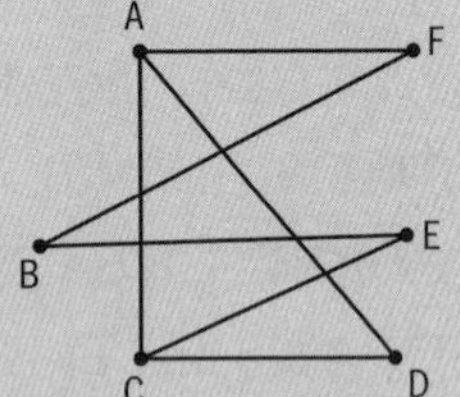

1 Has Hamiltonian paths. For example, BECDA is a Hamiltonian path. However, it is not possible to find a Hamiltonian cycle as vertex B has degree 1.	*Find a walk that includes all the vertices of the graph exactly once. To obtain a cycle you must return to the first vertex which is not possible in this case.*
2 Has Hamiltonian paths and cycles. For example, BFADCE is a Hamiltonian path and BFADCEB is a Hamiltonian cycle.	*Find a walk that includes all the vertices of the graph exactly once. To obtain a cycle, return to the initial vertex.*

Definition

A graph is said to be **Hamiltonian** if it has at least one Hamiltonian cycle. A graph is **semi-Hamiltonian** if it has at least one Hamiltonian path but no Hamiltonian cycle.

Example 8 shows that a graph may have Hamiltonian paths but not Hamiltonian cycles. Graph 2 is a Hamiltonian graph while graph 1 is semi-Hamiltonian.

Exercise 4F

1 In the following graphs, determine whether or not it is possible to find

i a Hamiltonian path

ii a Hamiltonian cycle.

a

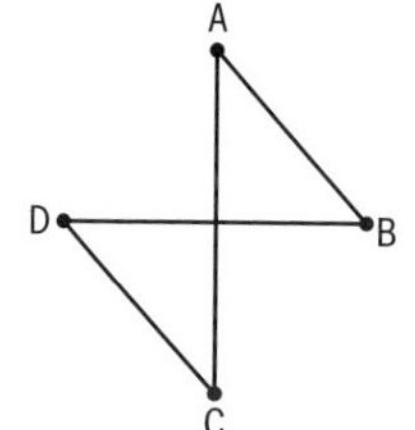

b

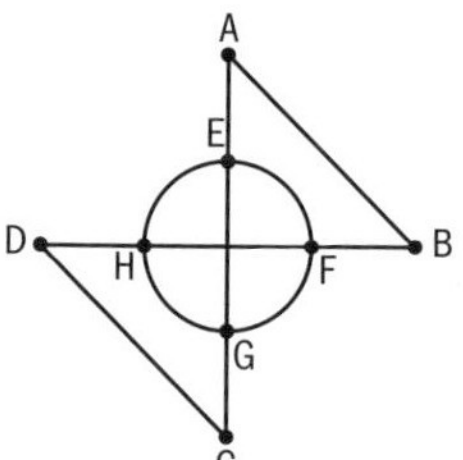

c

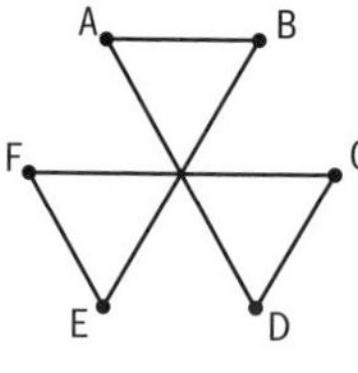

d

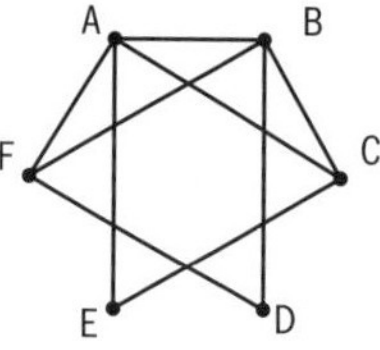

e

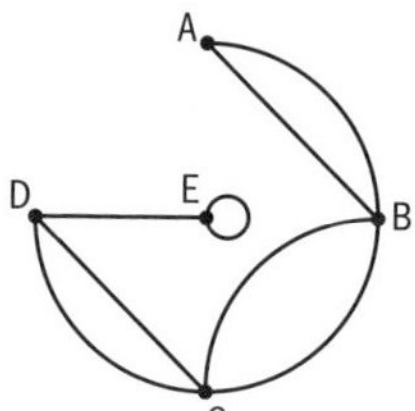

f

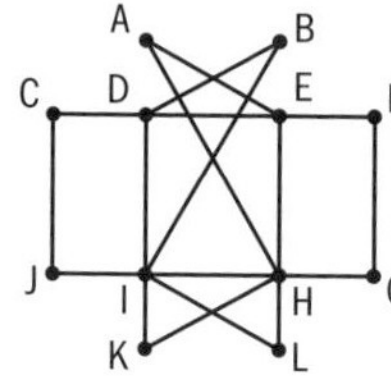

g

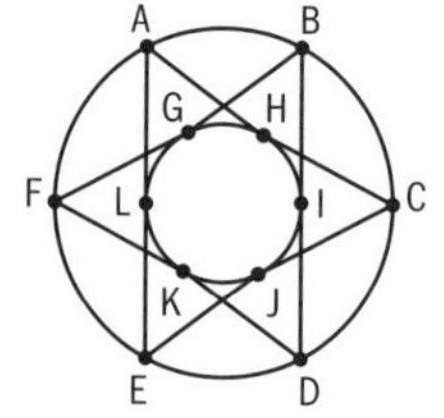

h

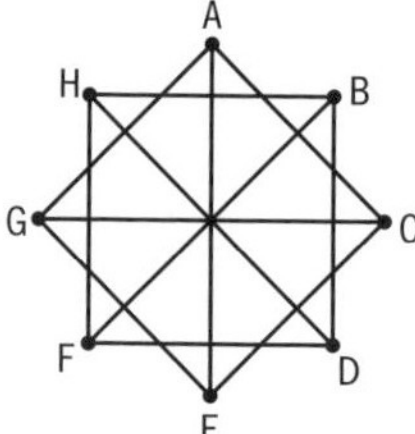

2 Draw a representation of K_5 and list all Hamiltonian cycles of this graph.

3 Prove that any bipartite graph containing an odd number of vertices cannot contain a Hamiltonian cycle.

4.6 Eulerian circuits and trails

In this last section of chapter 4 we are going to address the Königsberg bridge problem.

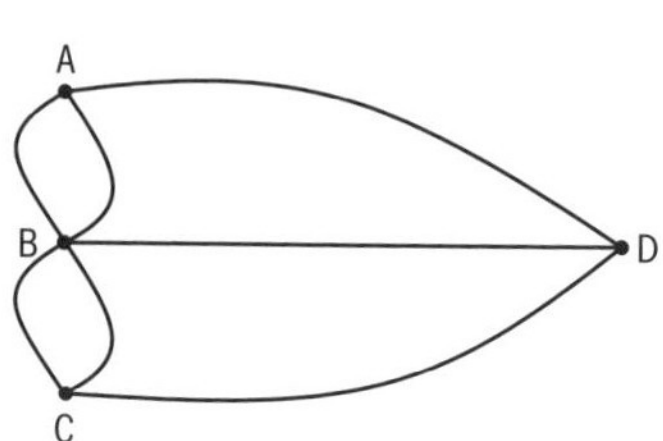

The residents of the city went for walks, trying to find a way to walk around the city in such a way as to cross each bridge only once and return to the starting point. (See page 103 for map.) Euler represented this situation by the multigraph shown here, where the city sections are represented by the vertices and the bridges are represented by the edges.

Euler proved that a solution to this problem was not possible. In order to prove this result we need to introduce a few new terms and theorems.

Let $G = (V, E)$ be a connected graph. A trail where each edge from E appears exactly once is called an **Eulerian trail**. A closed trail or circuit with the same property is called an **Eulerian circuit**.

Note that in a simple graph, every cycle is always a circuit. However a circuit may not be a cycle as it may go through the same vertex more than once, even if it does not cross the same edge twice.

Theorem 11

A connected graph contains an Eulerian circuit if and only if every vertex of the graph is of even degree.

Proof:

If the graph contains an Eulerian circuit we can arrive at and travel away from each vertex. Suppose that we start from one vertex and, as we follow the circuit from one vertex to another we erase each edge on which we travel. Every time we erase that edge, we reduce the degree of the two adjacent vertices by 1. Eventually we arrive at the starting vertex and there are no more edges; therefore all the vertices have degree zero. We conclude that the reduced degrees at each vertex must be multiples of 2 (once for each edge adjacent to it or for a loop that has a degree of 2), and therefore even.

Conversely, if we assume that each vertex has an even degree we can travel from and back to each vertex, therefore since the graph is connected we can form an Eulerian circuit. Q.E.D.

Corollary

Let $G = (V, E)$ be a connected graph. The graph G has an Eulerian trail if and only if it contains exactly two vertices of an odd degree.

Proof:

Suppose there is an Eulerian trail starting from the vertex A and finishing at the vertex B. Suppose that we follow this trail and erase each edge on which we travel. The first edge and the last edge contribute with only 1 to the degree of the vertices A and B, while all the other edges contribute with 2 to the degree of all intermediate vertices. We can conclude that A and B have odd degrees and all the remaining vertices have even degree.

Conversely, if we call the vertices with the odd degree A and B and connect them with additional edge AB, the added edge contributes with a degree 1 to the degree of both vertices. The new graph has all the vertices of even degree and therefore, by Theorem 11, it has an Eulerian circuit. If we walk from vertex A and bypass all but the added edge, we will finish at vertex B and we walk along an Eulerian trail. Q.E.D.

The previous theorems are the key to Euler's solution of the Königsberg bridges.

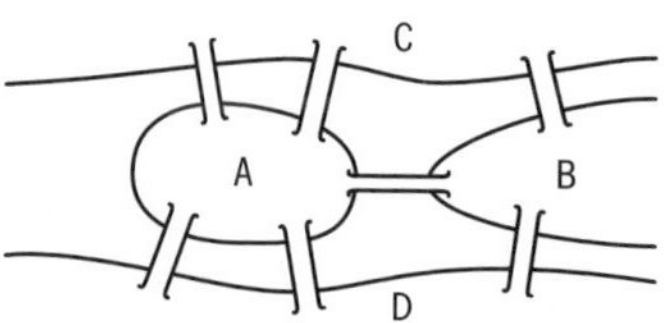

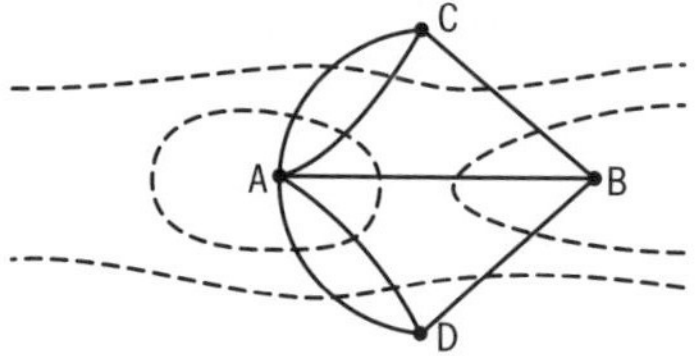

The degrees of the vertices are $deg(\text{A}) = 5$, and $deg(\text{B}) = deg(\text{C}) = deg(\text{D}) = 3$. Therefore by Theorem 11, no Eulerian circuit is possible in such a graph. So, the inhabitants of Königsberg could not walk in a circuit that included all the bridges.

Note that if we take out one bridge it would be possible to find an Eulerian trail, but still not an Eulerian circuit.

Königsberg was heavily bombed during the final weeks of World War II. Two of the seven original bridges did not survive the bombing. Two others were later demolished and replaced by a modern highway. The three other bridges remain, although only two of them are from Euler's time. Thus, as of 2000, there are now five bridges in Kaliningrad. The new graph has two vertices of degree 2, a vertex of degree 3 and a vertex of degree 5. Therefore, an Eulerian path is now possible.

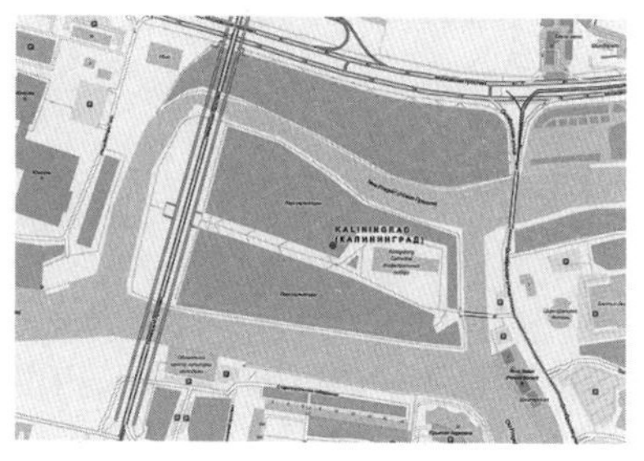

Example 9

State in which of these graphs it is possible to find an Eulerian trail and/or an Eulerian circuit. Provide an example in each case or give a reason why it cannot exist.

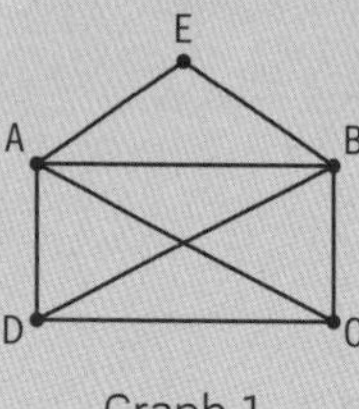

Graph 1

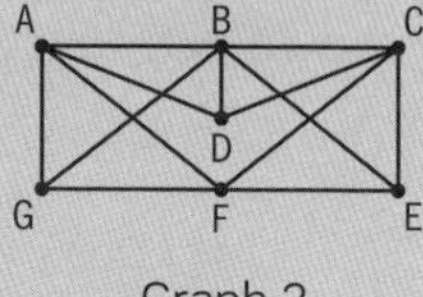

Graph 2

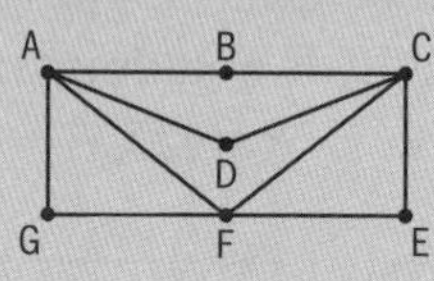

Graph 3

Graph 1 has Eulerian trails, for example: D→A→E→B →A→C→B→D→C; it is not possible to find an Eulerian circuit as it has two vertices of odd order.	*Start and finish with the vertices of odd order (C and D).*
Graph 2 is not Eulerian because the vertices B, D, E and G have odd order.	*Use the Corollary of Theorem 11.*
Graph 3 is Eulerian. For example A→B→C→D→A →G→F→E→C→F→A.	*Use Theorem 11.*

Exercise 4G

1 Determine whether or not in the following graphs it is possible to find

i an Eulerian trail   ii an Eulerian circuit.

a
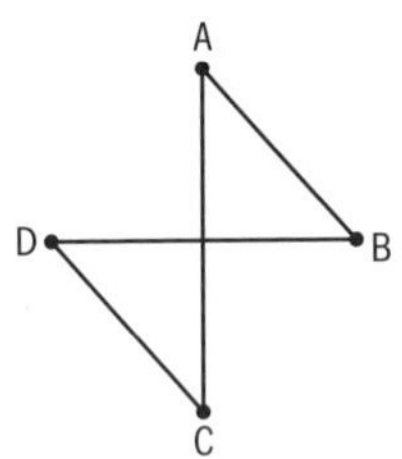

b
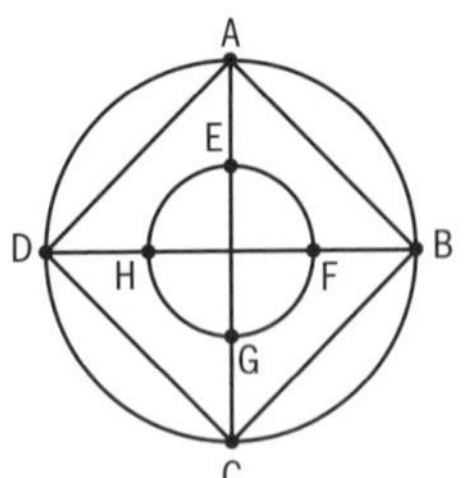

c
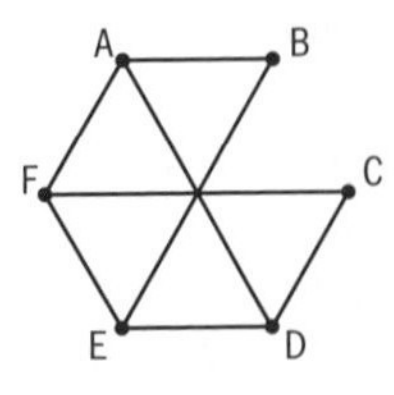

d
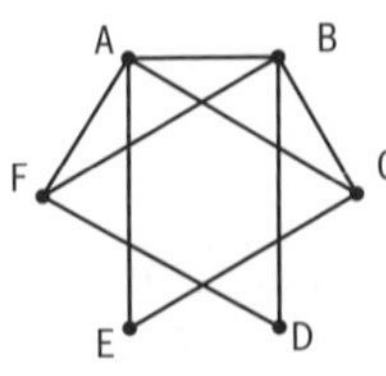

e
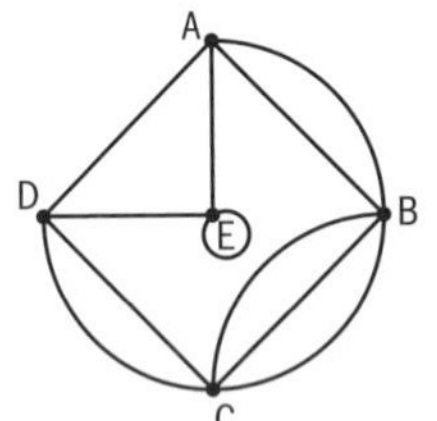

f
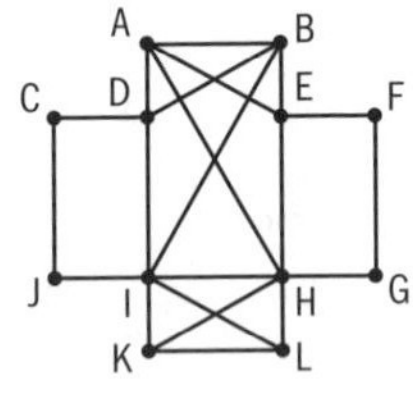

g
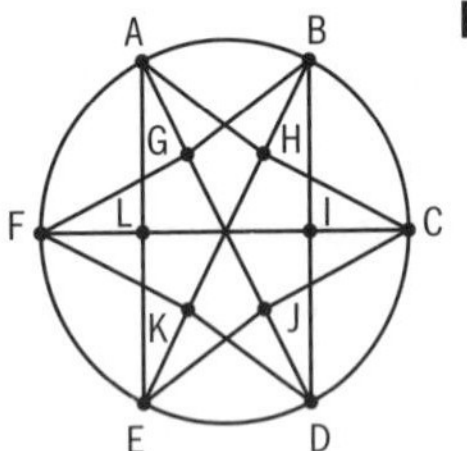

h
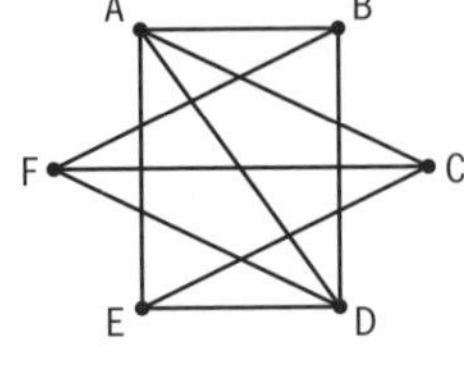

2 Given the complete graph K_4 and a path of length l between any two vertices in the graph, find the number of different paths when:

a $l = 2$ b $l = 3$.

3 Given the complete bipartite graph $K_{3,3}$ and a path of length l between any two non-adjacent vertices in the graph, find the number of different paths when:

a $l = 3$ b $l = 4$.

4 The floor plan of a certain shopping area is shown below. There are five rooms A, B, C, D, E and doorways are shown between the rooms and to the outside area O.

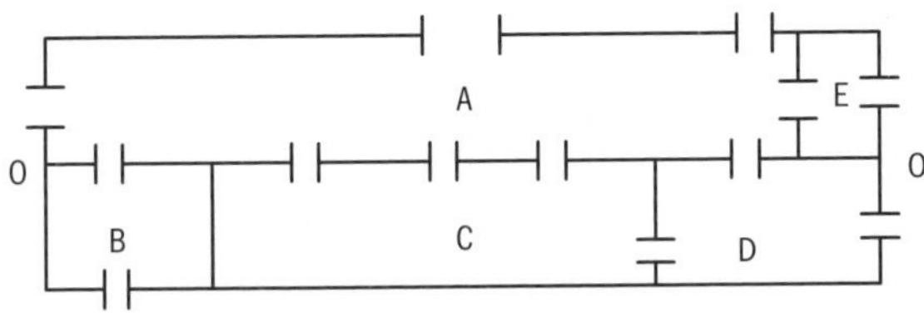

a Draw an appropriate graph representing the situation above where rooms will be associated to vertices and doorways between the rooms will be associated to edges between the corresponding vertices.

b Is it possible to enter the shopping area and pass through each doorway only once before you exit? Justify your answer.

c Is it possible to enter the shopping area and pass through each room only once before you exit? Justify your answer.

EXAM STYLE QUESTION

5 Consider the simple graphs G and H.

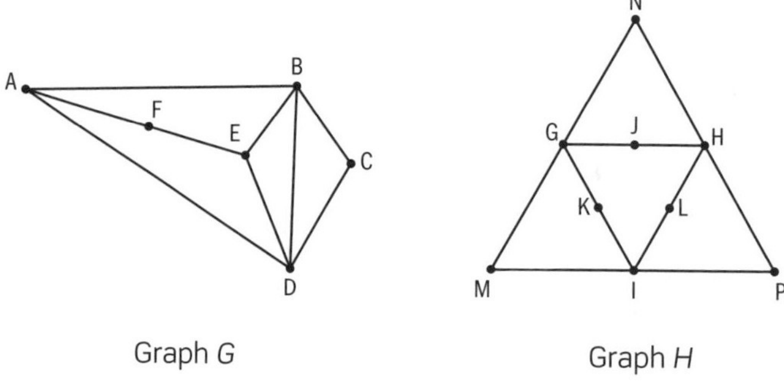

Graph G Graph H

a Show that graph G has an Eulerian trail starting with vertex E, stating clearly the sequence of vertices.

b Show that graph H has an Eulerian circuit.

c Decide whether or not G and H have Hamiltonian cycles.

6 Let G be a simple graph containing eight vertices.

a Show that G and the complementary graph G' cannot both contain an Eulerian trail.

b State with a reason whether or not the same is true for any simple graph containing an even number of vertices.

7 There are five Platonic solids: the tetrahedron, the cube, the octahedron, the dodecahedron and icosahedron.

a Draw the planar representation of each Platonic solid.

b Suppose that m is the number of edges that each region is bounded by, and n is the degree of each vertex. Use Euler's formula to show that each platonic solid must satisfy the inequality $(m - 2)(n - 2) < 4$.

c Hence show that there are exactly five Platonic solids.

Review exercise

1 Consider a group of 5 people that meet at a party. Is it possible for each of them to shake hands with:

a exactly 3 other people from the group

b exactly 4 other people from the group?

In each case, if possible, represent the solution in a form of a graph.

2 For each graph below, write down:

  a the number of vertices
  b the number of edges
  c the degree of each vertex
  d the adjacency table.

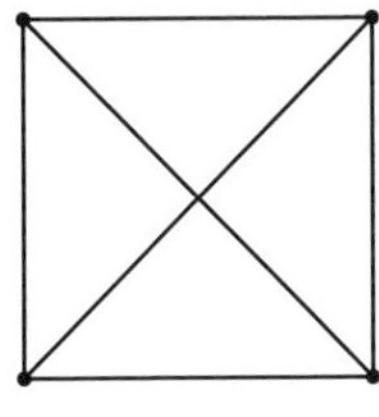

Graph G

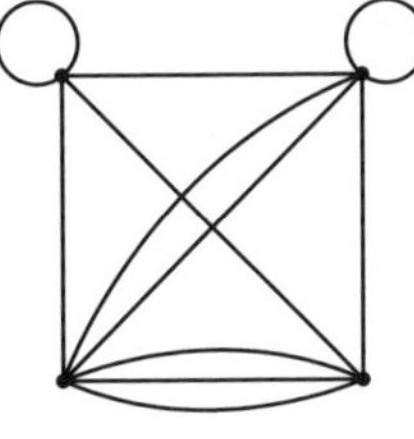

Graph H

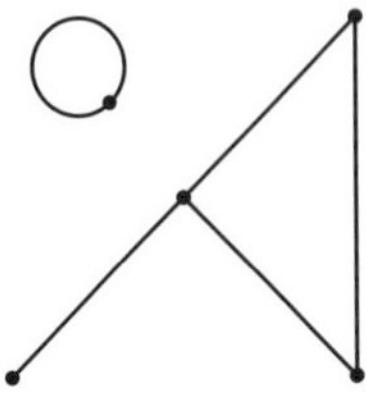

Graph J

3 A graph is called ***r*-regular** if all the vertices have the same degree r.

  a How many vertices does a 3-regular graph have if it has 12 edges?
  b Is it possible to have a regular simple graph with 14 edges? Explain your solution.
  c How many regular simple graphs are there with p edges, where p is a prime number?

4 Explain whether or not it is possible to have a cycle of odd length in a bipartite graph.

5 Prove that any subgraph of a bipartite graph must be bipartite.

6 Determine whether or not the following graphs are isomorphic. Explain your answer.

a 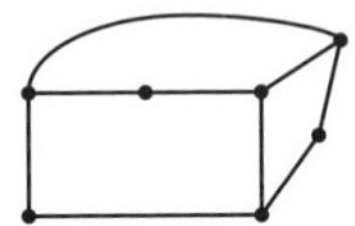

b 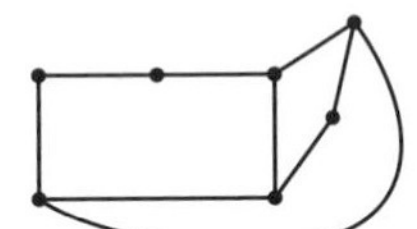

7 Given a complete graph K_5, find the number of trails between two of its vertices with length no longer than 3.

8 State, with reasons, which of these graphs are bipartite:

a 

b 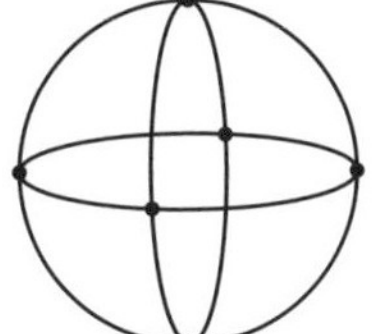

c 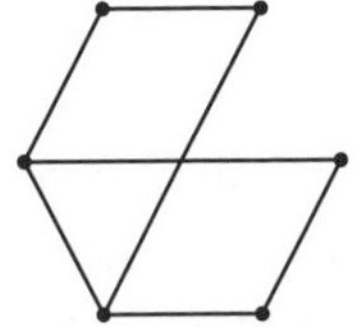

d 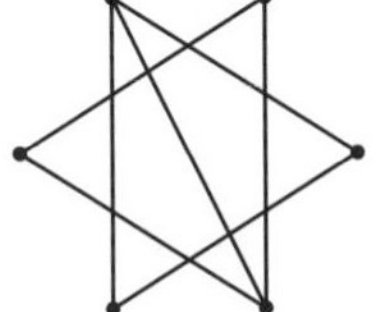

EXAM STYLE QUESTION

9 Prove that if a graph G with an odd number of vertices is k-regular then k is even.

10 A **cycle** C_n, $n \geq 3$, is a graph with n vertices of order of 2, i.e. a 2-regular graph with order n. A **wheel** W_n, $n \geq 3$, is a graph that consists of a cycle C_n and an additional point that is connected to all the vertices in the cycle. Here are some examples of cycles and wheels:

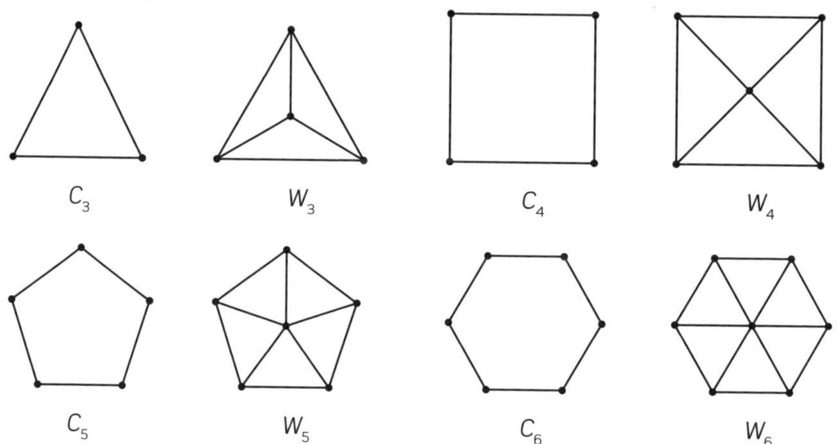

a Draw the complementary graph of C_5. Is the complementary graph isomorphic to the original graph? If yes, construct an isomorphism between the two graphs.

b Show that the number of edges in a wheel W_n is twice the number of edges in a cycle C_n.

c Are any of these graphs C_n or W_n isomorphic to a complete graph K_n?

11 Show that a cycle graph C_n, $n \geq 3$, is bipartite if and only if n is even.

12 Explain why no wheel graph W_n, $n \geq 3$, can be bipartite.

13 a Show that C_n, $n \geq 3$, contains an Eulerian circuit.

b Investigate whether or not W_n, $n \geq 3$, contains an Eulerian circuit or trail.

14 Determine whether or not the following graphs are planar.

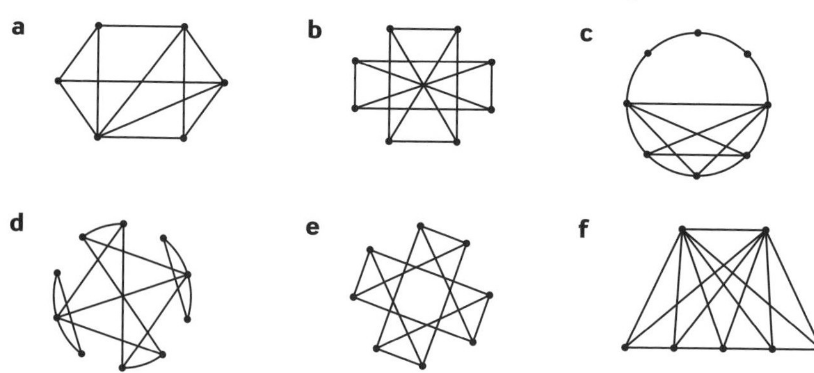

If the graph is planar, draw one of its planar embeddings; otherwise say why it cannot be planar.

15 Given a 6-regular graph with 10 vertices, find the number of regions the plane is divided into by a planar embedding of the graph.

16 Given that a connected planar graph has 20 edges and it divides the plane into 15 regions, find the number of vertices of the graph.

EXAM STYLE QUESTIONS

17 A canal system divides a town built near the mouth of a river into six land masses connected by fourteen bridges, as shown in the diagram.

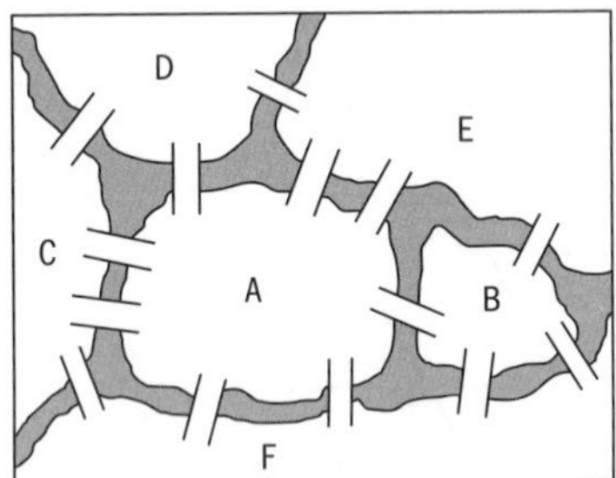

a Draw a planar graph to represent this map.

b Write down the adjacency table of the graph.

c List the degree of each of the vertices.

d State, with reasons, whether or not this graph has

i an Eulerian circuit **ii** an Eulerian trail.

18 The graph G and its complement G' are simple connected graphs, each having 12 vertices. Show that G and G' cannot both be planar.

19 A graph G has n vertices with degrees 1, 2, 3, …, n. Prove that $n \equiv 0 \pmod 4$ or $n \equiv 3 \pmod 4$.

20 a A connected planar graph G has e edges and v vertices.

i Prove that $e \geq v - 1$.

ii Prove that $e = v - 1$ if and only if G is a tree.

b A tree has n vertices of degree 1, two of degree 2, one of degree 3 and one of degree 4. Determine n and hence draw a tree that satisfies these conditions.

Chapter 4 summary

Definitions:

A graph $G = (V, E)$ consists of a set V of vertices and a set E of unordered pairs of vertices called edges. The number of vertices, $|V|$, is the **order** of G and the number of edges, $|E|$, is the **size** of G.

Two vertices A and B are said to be **adjacent** if they are joined by an edge. The number of edges incident in a vertex V is called the **degree** of the vertex and is denoted $deg(\text{V})$. Multigraphs are graphs that include multiedges and/or loops.

$H = (V', E')$ is called a **subgraph** of a graph $G = (V, E)$ if and only if $V' \subseteq V$ and $E' \subseteq E$ where $V' \neq \varnothing$, and the edges in E' are adjacent to the vertices in V'.

A **walk** is a sequence of linked edges. Usually we describe a walk by listing the vertices in order as we walk it. The **length of a walk** is its total number of edges listed.

A **path** is a walk with no repeated vertices.

The **degree sequence** of a path lists the degrees of the vertices in the order we pass through them as we walk along the path.

A **cycle** is a walk that begins and ends at the same vertex, and has no repeated vertices. The **length of a cycle** is its total number of distinct vertices listed.

A **trail** is a walk with no repeated edges.

A **circuit** is a walk that begins and ends at the same vertex, and has no repeated edges.

Theorem 1 (Handshaking lemma): if $G = (V, E)$ and $|E| = e$ then $\sum_{\text{A}\in V} deg(\text{A}) = 2e$.

Corollary: The number of vertices of odd degree in a graph $G = (V, E)$ is always even.

Classifications of graphs:

When the edges of a graph/multigraph are assigned a number – the weight – we obtain a weighted graph.

A **directed graph** or **digraph** is a graph where the edges have a direction associated with them. Digraphs can also be weighted graphs.

A **simple graph** is an unweighted, undirected graph containing no loops or multiple edges.

A graph is **connected** when there is a path from any vertex to any other vertex in the graph. A graph that is not connected is said to be disconnected.

Theorem 2: Let G be a simply connected graph of order n, for $n \geq 2$. Then G has at least two vertices with the same degree.

Definitions: A **tree** is a connected graph with no cycles. A **forest** is a graph without cycles. A subgraph T of a graph G is called a **spanning tree** if T is a tree and includes all the vertices of G.

Theorem 3: Let $G = (V, E)$ be a finite graph with more than one vertex. The following statements are equivalent:

1. G is a tree.
2. Each pair of vertices of G is connected by exactly one path.
3. If a is an edge of G, $G - \{a\}$ is disconnected.
4. G is cycle-free and has $n - 1$ edges.
5. G is connected and has $n - 1$ edges.

Definition: The graph K_n is called a **complete graph of order *n*** and it is characterized by the property 'each vertex is adjacent to every other vertex'.

Theorem 4: Let K_n be the complete graph of order n, $n \in \mathbb{Z}^+$. The number of edges (the size) of K_n is $\frac{n(n-1)}{2}$.

Definition: A simple graph $G = (V, E)$ is **bipartite** if its set of vertices V can be partitioned into two disjoint sets M and N such that each edge of G connects a vertex of M with a vertex of N.

Theorem 5: The number of edges of the complete bipartite graph $K_{m,n}$ is mn, i.e. $|K_{m,n}| = mn$.

Isomorphism invariants: If two graphs G and H are isomorphic, i.e. we can redraw one to obtain the other one, then:

- The size of G equals the size of H.
- The order of G equals the order of H.
- The sequence of the degrees of the vertices of G is a permutation of the sequence of the degrees of the vertices of H.
- The number of connected components of G and H is equal.
- The length of the cycles of G matches the lengths of the cycles of H.

Definitions: A graph G is **planar** if and only if we can draw it in a plane without any edges crossing over each other. H is called a **planar embedding** of G.
Cycles of planar graphs divide the plane into regions called **faces.** Graphs are called **regular** if all the vertices have the same degree.

Theorem 6: Given a connected planar graph G, the number of vertices v, edges e and faces f satisfy the formula: $v - e + f = 2$.

Corollary 1: Let G be a connected planar graph with at least 3 vertices. Then $e \leq 3v - 6$.

Corollary 2: Let G be a connected planar graph with no triangles. Then $e \leq 2v - 4$.

Theorem 7: K_5 is not planar.

Theorem 8: $K_{3,3}$ is not planar.

Theorem 9 (Kuratowski's theorem): A finite graph is planar if and only if it has no subgraph homeomorphic to K_5 or $K_{3,3}$.

Corollary: K_n is not planar for $n \geq 5$.

Definition: Let $G = (V, E)$ be a simple graph of order n. Consider the representation of K_n that has the vertices of V. Then the **complement** of G, denoted by G', is a graph that contains the same set of n vertices V as the graph G and contains all the edges of K_n that G does not contain.

Definitions: A graph is **Hamiltonian** when it is possible to walk around the graph, visit each vertex exactly once and return to the starting vertex, i.e. a closed path exists. A graph is **Eulerian** when it is possible to walk around the graph and cross each edge exactly once, i.e. a closed trail exists. A closed **Hamiltonian walk** is called a **Hamiltonian cycle**; a closed **Eulerian trail** is called an **Eulerian circuit**.

Theorem 10: Let $G = (V, E)$ be a simple connected graph. If $|V| = n, n \geq 3$, and for each vertex $V_i \in V$, $\deg(V_i) \geq \frac{n}{2}$ then the graph G has a Hamiltonian cycle.

Theorem 11: A connected graph contains an Eulerian circuit if and only if every vertex of the graph is of even degree.

Corollary: Let $G = (V, E)$ be a connected graph. The graph G has an Eulerian trail if and only if it contains exactly two vertices of an odd degree.

5 Applications of Graph Theory

CHAPTER OBJECTIVES:

10.9 Graph algorithms: Kruskal's and Dijkstra's.

10.10 Chinese postman problem. Travelling salesman problem. Nearest-neighbour algorithm for determining an upper bound. Deleted vertex algorithm for determining a lower bound.

Before you start

You should know how to:

1 Recognize planar, Eulerian and Hamiltonian graphs and know the terminology used to study them, e.g. draw this graph in planar form. State a Hamiltonian cycle and explain why this graph is not Eulerian.

A planar form of the graph is shown below right

ABCDEA is a Hamiltonian cycle as it goes through each vertex exactly once.

This graph has 2 vertices (B and D) with odd degree, therefore it is not Eulerian. It does contain an Eulerian trail, so it is semi-Eulerian.

2 Determine spanning trees of graphs, e.g. draw two spanning trees of the graph in question **1**. As there are 5 vertices the spanning trees will have 4 edges and contain no cycles.

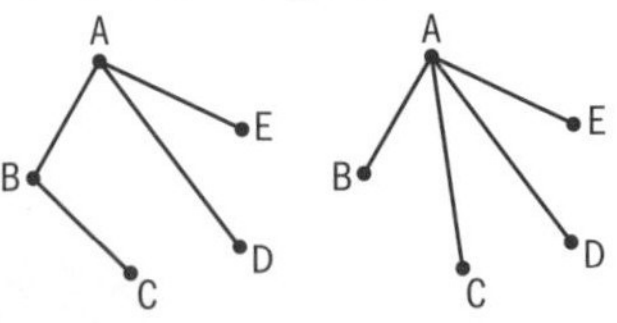

Skills check:

1 Draw the following graphs in planar form. State whether or not they are Hamiltonian and/or Eulerian.

a

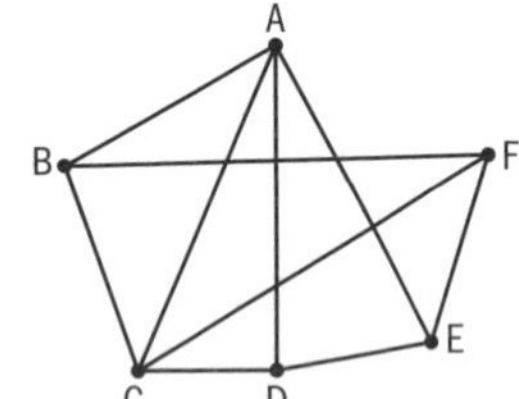

b

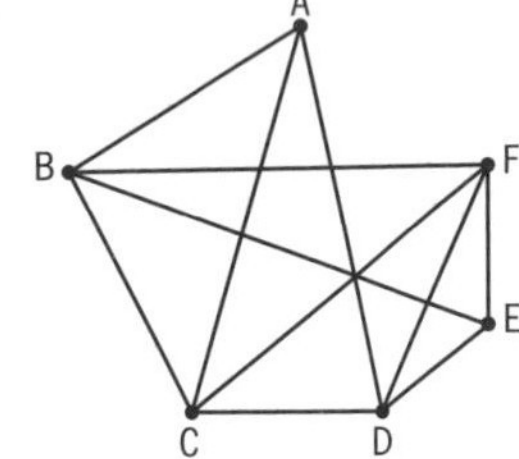

2 Consider the graph in **1b**. State, with reasons, the number of edges of a spanning tree of this graph. Draw five distinct spanning trees of this graph, including an example of a rooted tree.

Further algorithms and methods

The development of technology is closely related to the development of mathematical algorithms that allow machines to perform routine (as well as more complex) tasks involving automated decision making. Whenever you use a search engine to find information or use a GPS to find the quickest way home you are using efficient algorithms that analyze huge amounts of information and rank options according to pre-selected criteria.

Algorithms have therefore changed the way we live but also the way we do mathematics. The importance of this area of mathematics is growing fast, and even areas previously closed to empirical approach are being challenged – proofs using computers are still a cause of discussion in the mathematical community.

In this chapter we are going to explore challenging problems in Graph Theory and learn algorithms to tackle them. You will also discover limitations of these techniques and implications of their applications.

? The four-color theorem states that any map (in a plane) can be filled in using four-colours such that regions sharing a common boundary do not share the same colour. F. Guthrie was the first to conjecture this theorem in 1852. Fallacious proofs were given by several mathematicians. In 1977, Appel and Haken constructed a computer-assisted proof showing that four colours were sufficient. However, because part of the proof consisted of an exhaustive analysis of many discrete cases by a computer, some mathematicians do not accept it. However, no flaws have yet been found, so the proof appears to be valid.

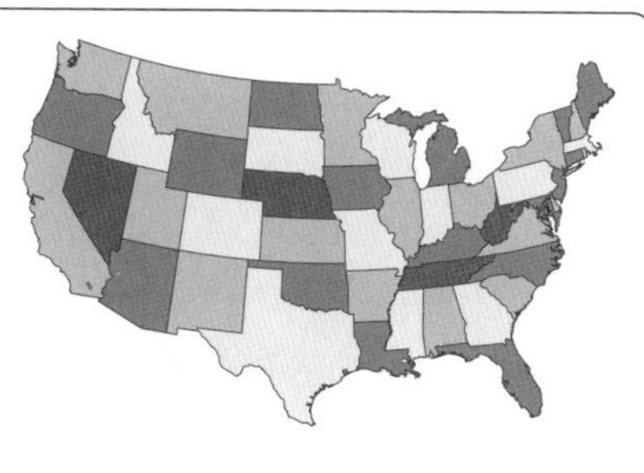

5.1 Graph algorithms: Kruskal's and Dijkstra's

Minimum Connector Problems

In this section we are going to look at effective ways of connecting all vertices of a given graph, i.e. we are going to find the **minimum spanning tree** of a given connected weighted graph. We start our study with a 'greedy algorithm' called **Kruskal's algorithm**.

Kruskal's algorithm first orders the edges of a graph by weight and then proceeds through the ordered list adding an edge to the partial minimum spanning tree, provided that adding the new edge does not create a cycle.

A **greedy algorithm** makes the locally optimal choice at each stage with the hope of finding a global optimum. However, in many problems a greedy strategy does not produce an optimal solution, but nonetheless a greedy strategy may yield locally optimal solutions that approximate a globally optimal solution in a reasonable time.

Joseph Bernard Kruskal, Jr. (1928–2010) was an American mathematician, statistician, and computer scientist. In Graph Theory, his best known work is Kruskal's algorithm for computing the minimal spanning tree of a weighted graph.

Let's use Kruskal's algorithm to find a minimum spanning tree, let's say T, of a weighted graph G with n vertices:

1. Select an edge with minimum weight to be the first edge of T.
2. Consider the weighted edges of $G - T$ which do not form a cycle with the already chosen edges of T. Pick the one with minimum weight and add the new edge and vertex to T (in case there is more than one with minimum weight, choose any of them).
3. Repeat step **2** until $n - 1$ edges have been chosen.

Example 1

Apply Kruskal's algorithm and find the minimum spanning tree of this graph.

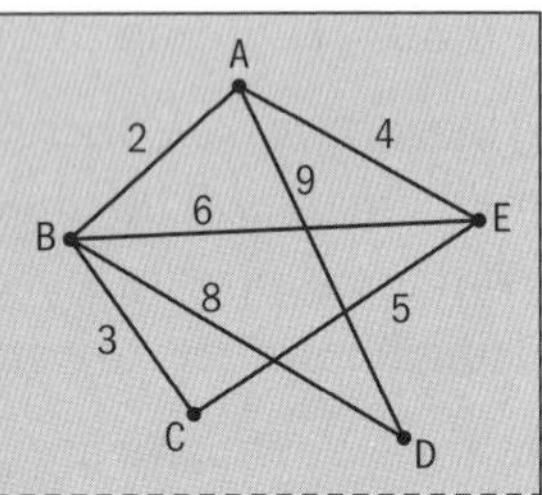

Start the tree T with edge AB with weight 2;

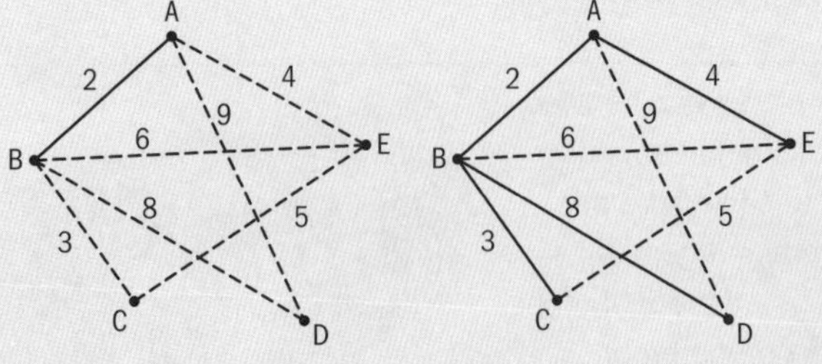

add the edge BC, AE and BD.
The minimum spanning tree has total weight $2 + 3 + 4 + 8 = 17$.

As the graph has 5 vertices, the tree will have 4 edges.

1. *Select the edge with minimum weight to be the first edge of T.*
2. *Consider the weighted edges of $G - \{AB\}$ which do not form a cycle with the already chosen edges of T. Pick the one with minimum weight and add the new edge and vertex to T.*
3. *Repeat until T has 4 edges. Add the weights of the edges of the tree to obtain its total weight.*

To make the application of this algorithm efficient when dealing with large graphs we are going to use special adjacency tables – **cost adjacency tables** – where the entries represent the weight of the edges. For example, the graph below shows the current network of streets between several villages where the weights represent the distance in kilometres.

When dealing with weighted graphs the adjacency tables are always cost adjacency tables. For simplicity in this chapter we will just call them adjacency tables.

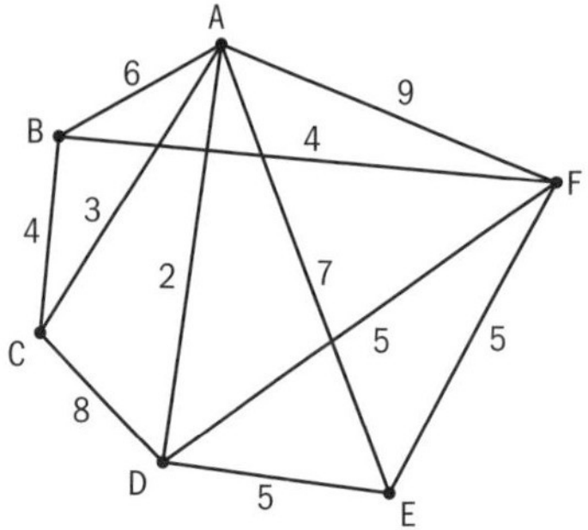

The adjacency table for this graph is:

	A	B	C	D	E	F
A	–	6	3	2	7	9
B	6	–	4	–	–	4
C	3	4	–	8	–	–
D	2	–	8	–	5	5
E	7	–	–	5	–	5
F	9	4	–	5	5	–

Suppose that a bike path is to be added to some of these streets in such way that:

- any two villages are connected
- the total length of the path is the shortest possible.

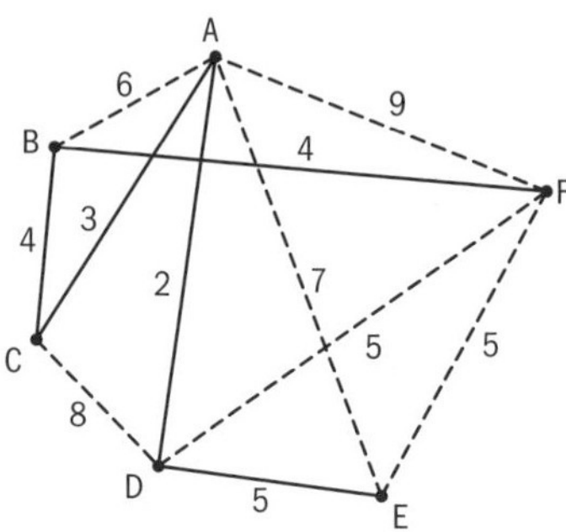

This is a real life example where a minimum spanning tree is the solution to the problem: a solution is the tree with edges AD, AC, BC, BF and DE with total weight 18, i.e. the total length of the path is 18 km (diagram, above right).

Note that this is not the only optimal solution – another solution could have been obtained if we chose the edge EF instead of DE (diagram, below right).

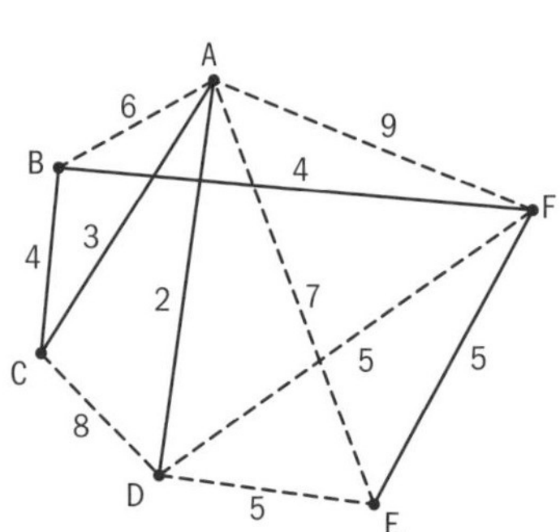

Example 2 shows how to apply Kruskal's algorithm to find a minimum spanning tree using an adjacency table.

Example 2

Apply Kruskal's algorithm to find the minimum spanning tree of the graph G with this adjacency table. State the total weight of the spanning tree.

	A	B	C	D	E	F
A	-	15	13	10	7	9
B	15	-	11	-	-	4
C	13	11	-	6	-	-
D	10	-	6	-	5	9
E	7	-	-	5	-	8
F	9	4	-	9	8	-

Start the tree T with edge BF with weight 4;

	A	B	C	D	E	F
A	-	15	13	10	7	9
B	15	-	11	-	-	4
C	13	11	-	6	-	-
D	10	-	6	-	5	9
E	7	-	-	5	-	8
F	9	4	-	9	8	-

1 Select an edge with minimum weight to be the first edge of T – this edge corresponds to the lowest non-zero entry on the table.

2 Consider the weighted edges of $G - \{BF\}$ which do not form a cycle with the already chosen edges of T. Pick the one with minimum weight and add the new edge and vertex to T.

3 Repeat until T has 5 edges. As the graph has 6 vertices, the tree will have 5 edges.

add the edge DE,

	A	B	C	D	E	F
A	-	15	13	10	7	9
B	15	-	11	-	-	4
C	13	11	-	6	-	-
D	10	-	6	-	5	9
E	7	-	-	5	-	8
F	9	4	-	9	8	-

and then CD, AE and EF.

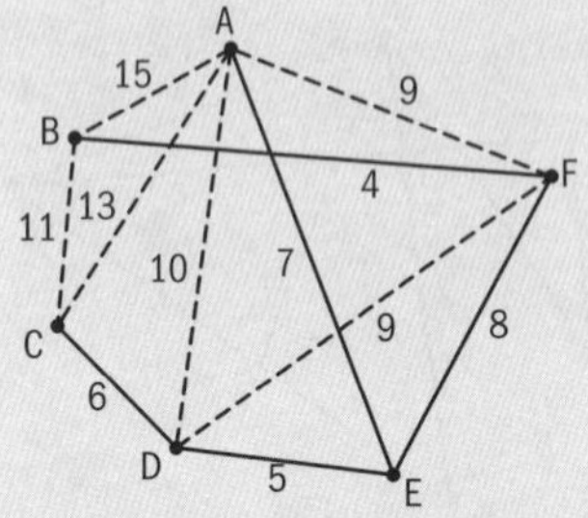

Watch for cycles being formed by drawing a sketch of the tree T.

The total weight of the spanning tree is $4 + 5 + 6 + 7 + 8 = 30$.

Add the weights of the edges of the tree to obtain its total weight.

Exercise 5A

1 Construct the adjacency table for each of the weighted graphs.

a

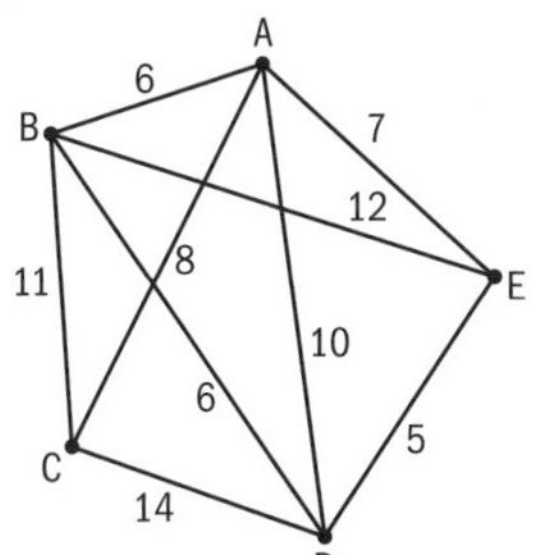

b

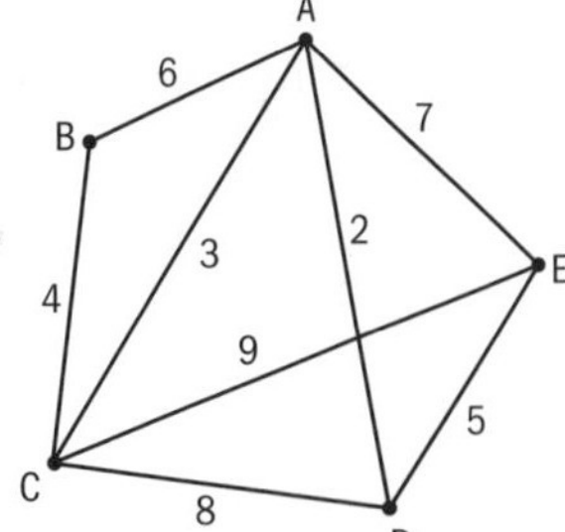

c

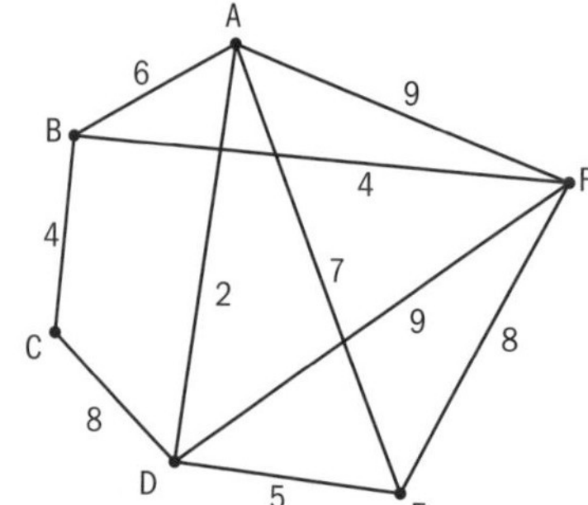

d

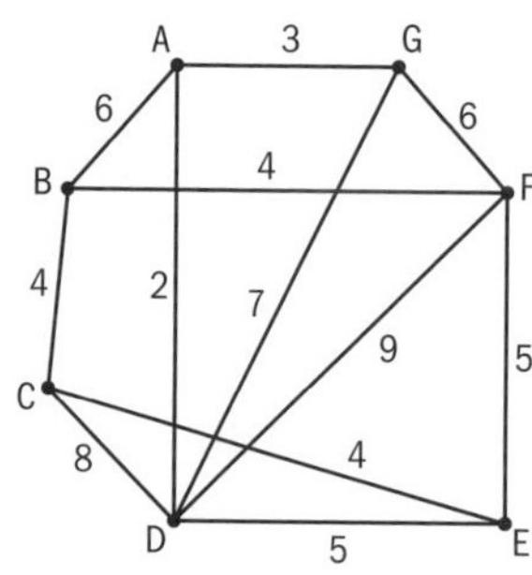

2 For each of the graphs in question **1**, apply Kruskal's algorithm and find a minimum spanning tree for the graph. In each case state the total weight of the tree.

3 For each of the adjacency tables below:

i draw the corresponding weighted graph, if possible in planar form.

ii apply Kruskal's algorithm and find a minimum spanning tree for each graph.

iii find the total weight of the minimum spanning tree found in **ii**.

a

	A	B	C	D	E
A	-	5	-	1	-
B	5	-	4	2	3
C	-	4	-	6	-
D	1	2	6	-	3
E	-	3	-	3	-

b

	A	B	C	D	E
A	-	5	-	1	-
B	5	-	4	-	3
C	-	4	-	2	-
D	1	-	2	-	3
E	-	3	-	3	-

c

	A	B	C	D	E	F
A	-	6	3	7	-	-
B	6	-	4	2	-	5
C	3	4	-	5	-	4
D	7	2	5	-	5	-
E	-	-	-	5	-	8
F	-	5	4	-	8	-

d

	A	B	C	D	E	F
A	-	6	3	10	-	7
B	6	-	4	2	-	5
C	3	4	-	5	-	-
D	10	2	5	-	5	9
E	-	-	-	5	-	8
F	7	5	-	9	8	-

4 Show that the graph given by the adjacency table in:

a **3a** is Eulerian. Hence state the length of the Eulerian circuit.

b **3b** is semi-Eulerian. Hence state the length of the Eulerian trail.

5 In a weighted graph G with 8 vertices, all the edges have different weights and all these weights are positive integers. Suppose that 5 is the weight of the edge with least weight in G. Find the least possible value for the total weight of a minimum spanning tree of G. Explain your reasoning.

6 In a weighted graph G with 10 vertices, all the edges have different weights. Let x be the weight of the edge with least weight in G. Given that a minimum spanning tree of G has a total weight of 90, find the maximum possible value of x. Explain your reasoning.

7 Joerg needs to install sockets that will be connected by an optical cable in his apartment so that he can watch TV, and use the phone and internet in all rooms. The positions of the sockets are shown on the graph. The distances between the sockets are given in metres.

Given that the cost of optical cable is €1.20 per metre, find the minimum price Joerg will pay for buying the cable.

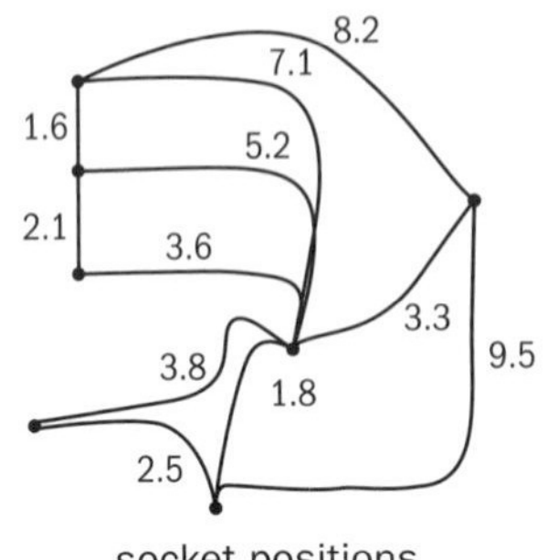

socket positions

8 Katharina plays a computer game in which she must visit rooms A, B, C, D, E, F, G, H and I in any order, and in each room she collects 10 points. The times, in minutes, between any two rooms in the first level of the game are given in the adjacency table shown. In order to advance to the next level of the game she must visit all the rooms in the shortest possible time.

Decide whether or not Kruskal's algorithm allows Katharina to find the minimum possible time that it may take her to visit all the rooms at the first level. Hence state where she should start if she wants to score at least 50 points within 10 minutes.

	A	B	C	D	E	F	G	H	I
A	0	2	3	4	0	0	2	0	0
B	2	0	3	2	3	0	0	0	0
C	3	3	0	0	0	4	0	0	3
D	4	2	0	0	2	0	4	0	0
E	0	3	0	2	0	5	0	4	0
F	0	0	4	0	5	0	0	0	2
G	2	0	0	4	0	0	0	6	0
H	0	0	0	0	4	0	6	0	5
I	0	0	3	0	0	2	0	5	0

Shortest Path Problems

Given a connected, weighted (and possibly) directed graph G, we are going to find the path between two given vertices which has the least possible weight.

A **shortest path tree**, in Graph Theory, is a subgraph of a given (weighted) graph constructed so that the distance between a selected root vertex and all other vertices is minimal.

A **search algorithm** is a procedure to find an item with specified properties among a collection of items. The items may be stored individually as data or may be elements of a search space defined by a mathematical formula or procedure.

? **Dijkstra's Algorithm**, conceived by Dutch computer scientist Edsger Dijkstra in 1959, is a **graph search algorithm** that solves the shortest path problem for a graph with nonnegative edge path costs (i.e. all the edges have nonnegative weights), producing a shortest path tree.

Dijkstra's Algorithm

Dijkstra's Algorithm assigns initial distance values and tries to improve them step-by-step.

Start with an **initial vertex**. Let the distance of a vertex be the distance from the initial vertex to it.

1 Assign to every vertex a distance value: zero for the initial vertex and infinity for all other vertices.
2 Mark all vertices as unvisited. Set initial vertex as current.
3 For the current vertex, consider all its unvisited neighbours and calculate the distance to each one of them. If this distance to a neighboring vertex is less than the previously recorded distance then overwrite this distance.
4 After considering all neighbours of the current vertex, mark it as visited. A visited vertex will not be checked ever again; its distance recorded now is final and minimal.
5 Set the unvisited vertex with the smallest distance (from the initial vertex) as the next "current vertex" and continue from step **3**.
6 When all the vertices have been visited, STOP.

The following example shows how to apply Dijkstra's Algorithm to a weighted but undirected graph.

Example 3

Use Dijkstra's Algorithm to find the shortest path between vertex B and each other vertex. Show all the steps of the algorithm and draw the solution paths.

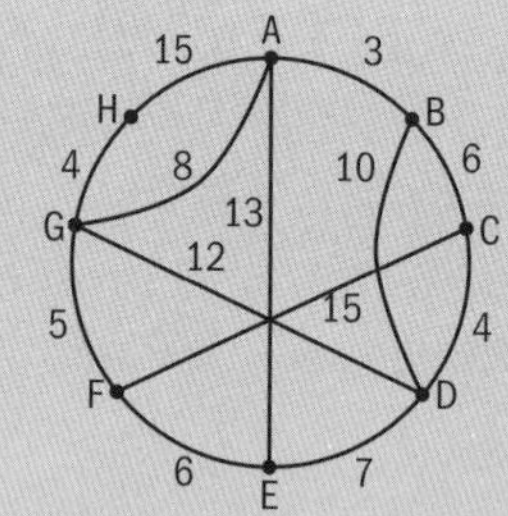

Step	A	B	C	D	E	F	G	H	vertex added
1	3(B)	0	6(B)	10(B)	∞	∞	∞	∞	B
2	...	...	6(B)	10(B)	16(A)	∞	11(A)	18(A)	A
3	...	...	...	10(B)	16(A)	21(C)	11(A)	18(A)	C
4	...	...	...	...	16(A)	21(C)	11(A)	18(A)	D
5	...	...	...	...	16(A)	16(G)	...	15(G)	G
6	...	...	...	...	16(A)	16(G)	...	...	H
7	...	...	...	...	16(A)	...	...	...	F
8	...	...	...	...	...	...	...	...	E

Apply Dijkstra's Algorithm starting from vertex B. State clearly which vertex you add at each step, its final distance to initial vertex B and mark the corresponding column on the table as inspected.

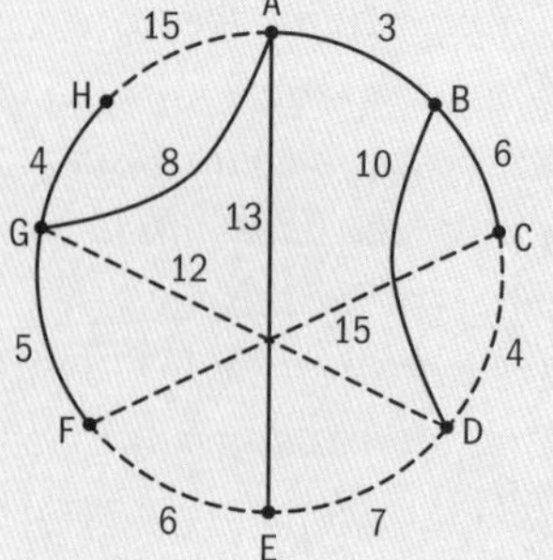

Draw the minimum connector tree from vertex B to each vertex; trace back from each vertex the path to the initial vertex using the information on the shaded entries of the table.

Limitation of Dijkstra's Algorithm

Dijkstra's Algorithm cannot be used if any weights are negative, as labels may become permanent before the 'cheapest' route is considered.

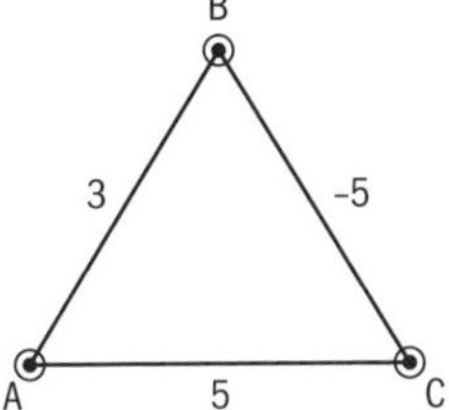

For example, ACB is clearly the shortest route from A to B but it cannot be found with Dijkstra's Algorithm.

The following example shows you an application of Dijkstra's Algorithm in context.

> There are different methods of recording the information needed to apply Dijkstra's Algorithm. However, despite apparent differences all the methods have in common the following necessary aspects:
>
> - Starting vertex
> - Total distance from each vertex to the starting vertex at each step
> - Ordered record of the vertices already inspected and their final distance to starting vertex.

Example 4

The graph shows the cheapest prices, in euros, of flights between several European cities on a given day. Use Dijkstra's Algorithm to find the cheapest route between Vienna and each other city on that day. Show all the steps of the algorithm and draw the solution routes.

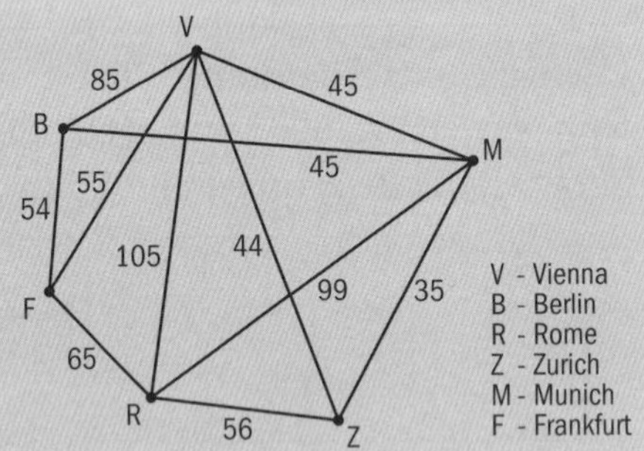

Step	V	B	F	R	Z	M	vertex added
1	0	85(V)	55(V)	105(V)	44(V)	45(V)	V
2	...	85(V)	55(V)	100(Z)	...	45(V)	Z
3	...	85(V)	55(V)	100(Z)	...	...	M
4	...	85(V)	...	100(Z)	...	...	F
5	...	...	...	100(Z)	...	...	B
6	...	...	...	...	...	...	R

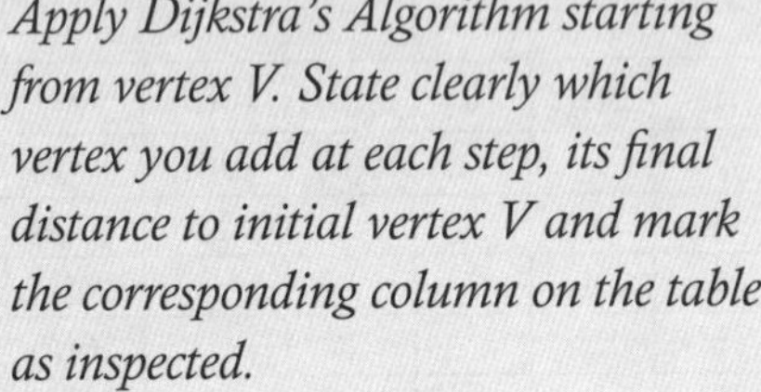

Apply Dijkstra's Algorithm starting from vertex V. State clearly which vertex you add at each step, its final distance to initial vertex V and mark the corresponding column on the table as inspected.

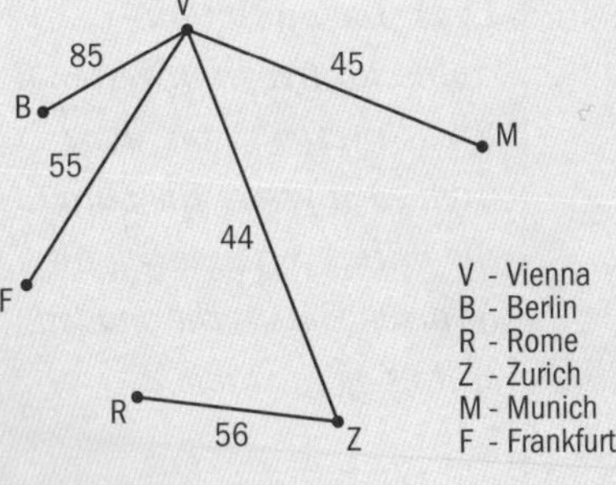

Draw the minimum connector tree from vertex V to each vertex; trace back from each vertex the path to the initial vertex using the information on the shaded entries of the table.
For example, the cheapest route from Vienna to Rome on that day is via Zurich. For all other cities, the direct flights are the best deals.

Exercise 5B

1 The graph G has this adjacency table.

	A	B	C	D	E	F
A	–	3	–	–	–	9
B	3	–	7	–	–	4
C	–	7	–	8	4	3
D	–	–	8	–	2	–
E	–	–	4	2	–	8
F	9	4	3	–	8	–

a Draw G in planar form.

b Use Dijkstra's Algorithm to find the shortest path between vertices A and D. Show all the steps in the algorithm and state the length of the shortest path.

2 Kristian is coordinating a project to design a road system to connect six towns, A, B, C, D, E and F. The possible roads and the construction costs, in hundreds of millions of euros, are shown in the graph to the right. Each vertex represents a town, each edge represents a road and the weight of each edge is the cost of building that road. Kristian needs to design the lowest cost road system that will connect the six towns.

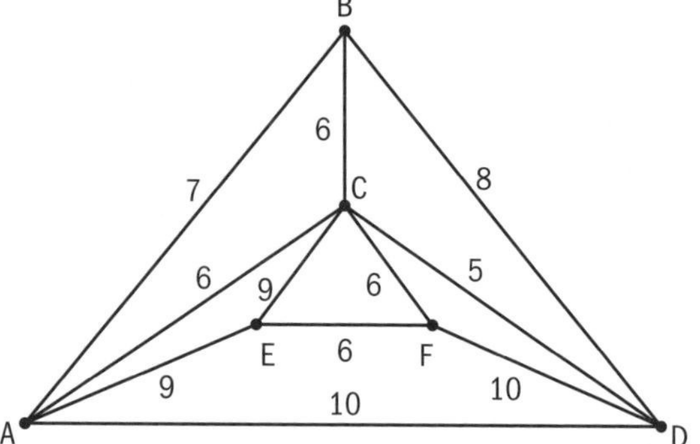

a State the name of an algorithm that allows Kristian to find the lowest cost road system.

b Find the lowest cost road system and state the cost of building it. Show clearly the steps of the algorithm.

3 The diagram here shows the weighted graph G.

a Write down the cost adjacency table for G.

b Use Kruskal's algorithm to find and draw the minimum spanning tree for G. Your solution should clearly indicate the way in which the tree is constructed.

c State with reasons whether or not the graph G

i is bipartite **ii** is semi-Eulerian.

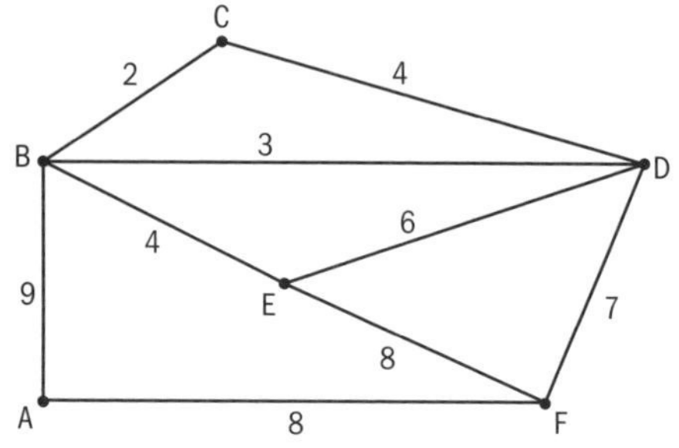

5.2 Chinese postman problem

In 1962, a Chinese mathematician called Kuan Mei-Ko was interested in a postman delivering mail to a number of streets in a town such that the total distance walked by the postman was as short as possible.

How can a postman in general ensure that the distance walked is a minimum?

In order to accomplish this task in an efficient manner, the postman would ideally choose a route that would allow him to avoid walking the same street more than once.

If we represent the town by a simple graph G whose edges represent the streets where the postman must deliver the mail, then the postman problem becomes the problem of finding an Eulerian trail. If the postman is to end his route at the starting point, the solution must be an Eulerian circuit. As an Eulerian circuit contains every edge of the graph exactly once, in the case of a weighted graph, the total weight of the circuit is the sum of the weight of all the edges.

Eulerian graphs and the terminology associated to them were studied in chapter 4.

Example 5

Consider the graph G here where each vertex represents a village that a postman needs to visit, and each edge is a road connecting two villages. The weights of the edges represent the distances between the corresponding villages.

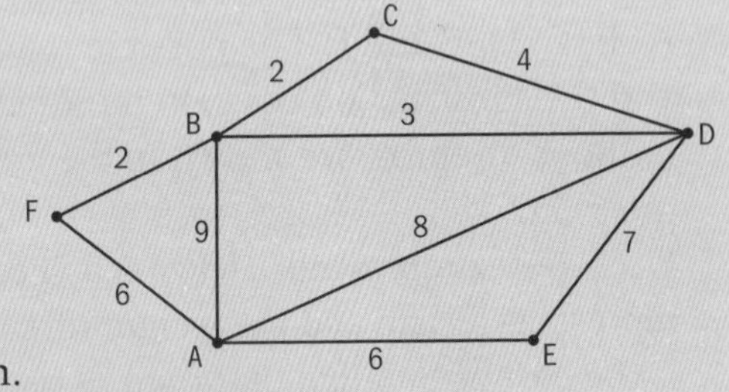

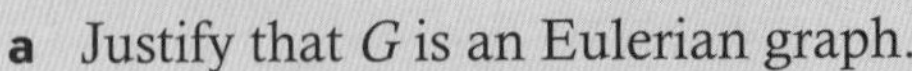

a Justify that G is an Eulerian graph.

b Find an Eulerian circuit and state its total length.

a All the vertices of G have even degree. Therefore the graph is Eulerian.	*deg(A) = deg(B) = deg(D) = 4* *deg(C) = deg(E) = deg(F) = 2*
b ABDAFBCDEA is an example of an Eulerian circuit of G. Its weight is 47.	*Start at any vertex and list the vertices in order as you move around the graph without crossing the same edge twice. Add the weights of all the edges to obtain the total weight of the circuit.*

The Chinese postman problem is an example of a routing problem. Routing is the process of selecting best paths in a network. Routing is performed for many kinds of networks, including the telephone network, electronic data networks such as the internet, but also transportation networks.

Now we are going to explore this problem further and find closed trails of minimum weight containing every edge of a graph G that may contain some vertices of odd order.

To find a minimum Chinese postman route we must walk along each edge at least once; we must also walk along the least pairings of odd vertices on one extra occasion.

Chinese postman algorithm

1 List all odd vertices.

2 List all possible pairings of odd vertices.

3 For each pairing, find the edges that connect the vertices with the minimum weight.

4 Find the pairings such that the sum of the weights is minimized.

5 On the original graph, add the edges that have been found in step **4**.

6 The length of an optimal Chinese postman route is the sum of all weights of the edges including the ones added to the total in step **4**.

7 Draw the route corresponding to this minimum weight.

IB exam questions will not include cases with more than four vertices of odd degree.

Example 6

Apply the Chinese postman algorithm to find the least weight closed trail containing every edge of this graph.

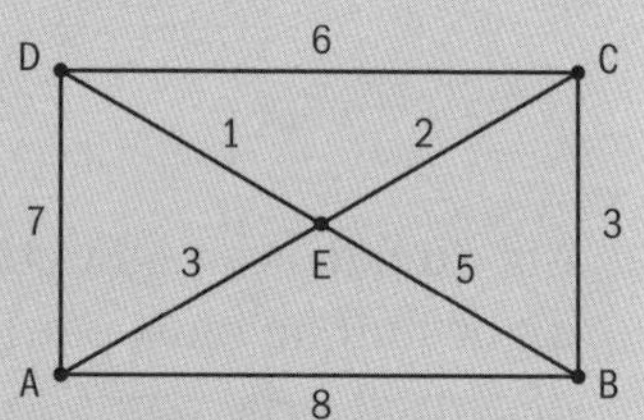

A, B, C and D are odd vertices.	*Step 1: List all odd vertices.*
The minimum weight connecting paths are: AB 8, BC 3, AC 5, BD 6, AD 4, and CD 3. Possible pairs where all odd vertices are connected are: AB and CD: 8 + 3 = 11, AC and BD: 5 + 6 = 11, AD and BC: 4 + 3 = 7.	*Step 2: List all possible pairings of odd vertices.* *Step 3: For each pairing find the edges that connect the vertices with the minimum weight.*
Add the edges AED and BC to obtain an Eulerian graph.	*Step 4: Find the pairings such that the sum of the weights is minimised.* *Step 5: On the original graph add the edges that have been found in Step 4.*
A minimum weight trail must have total weight 7 + 8 + 3 + 5 + 1 + 2 + 6 + 3 + 1 + 3 + 3 = 42	*Step 6: The length of an optimal Chinese postman route is the sum of the weights of all the edges.*
For example: AEABCBECDEDA 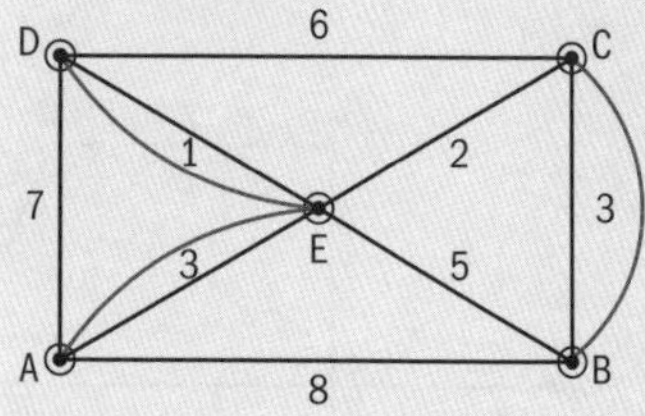	*Step 7: Draw the route corresponding to this minimum weight.*

Exercise 5C

1 For each of the following graphs:

a Write down the vertices of odd degree.

b Explain why it is impossible to walk along each edge exactly once and return to the starting vertex.

c Use the Chinese postman algorithm to find the route with minimum total weight that allows you to visit each edge at least once and return to the starting vertex.

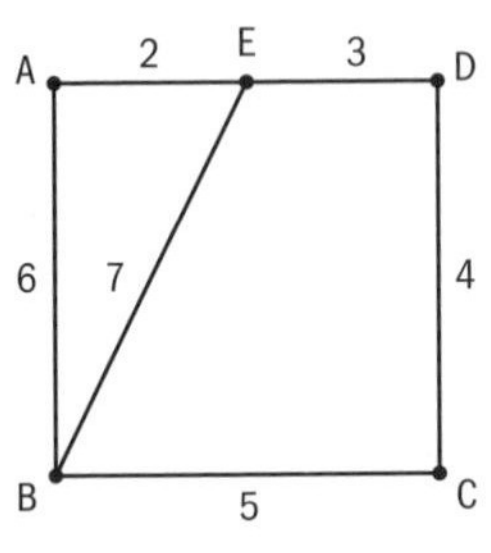

Graph *R*

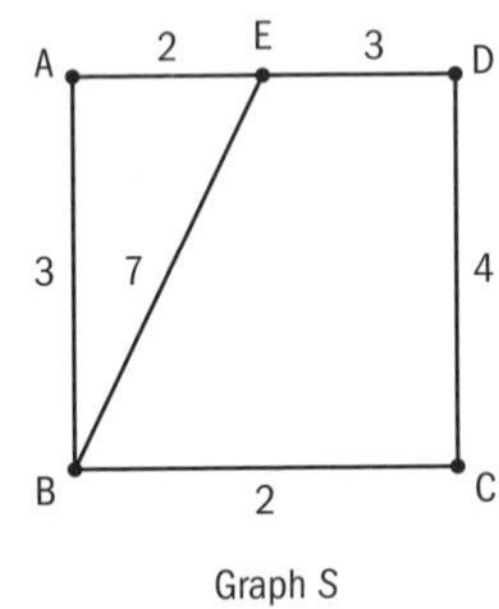

Graph *S*

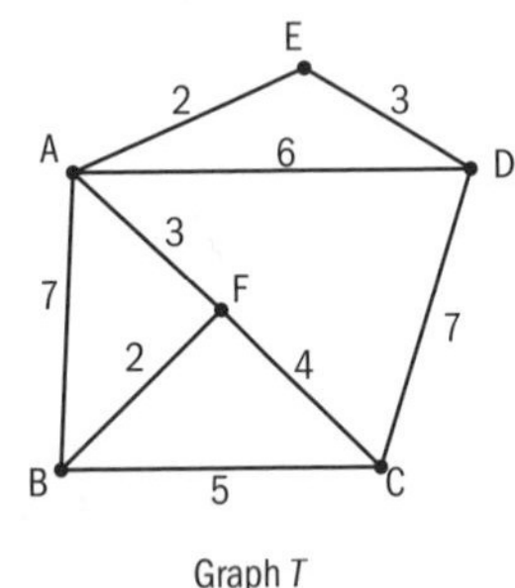

Graph *T*

2 Mr. Atente is a night guard and during the night shift he must patrol every single street of a residential complex. The plan of the streets and the time needed to patrol each street in minutes is shown in the diagram. State, with reasons, if it is possible for Mr. Atente to patrol the whole complex during his night shift from 10 p.m. until 6 a.m. If it is possible, then state how many minutes he will have for a break. If not possible, state how much longer he would need to stay in order to fulfil his duty.

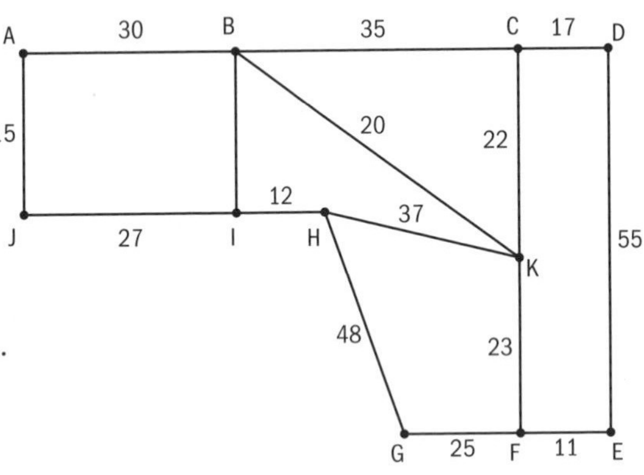

3 A snow-plough must drive along all the main roads shown on the graph, starting and finishing at the garage at A. The distances shown represent kilometres.

a Show that the snow-plough must drive at least 25 km to clean all the roads shown.

b Find the least distance it must actually travel, showing clearly a possible route.

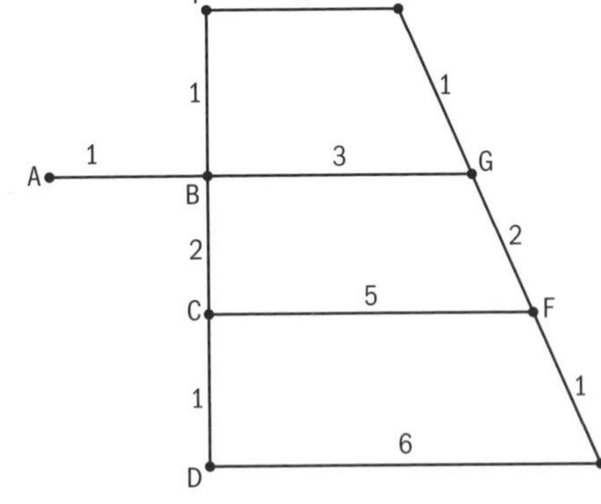

5.3 Travelling Salesman Problem

To conclude our study of Graph Theory algorithms we are going to look at the challenging problem of finding a tour for a salesman that visits every city (vertex) of a list (graph) exactly once.

Meanwhile, the Hamiltonian problem evolved to a more sophisticated and complex problem: the problem of finding a Hamiltonian cycle of least total weight in a weighted graph G. Such is the nature of the famous Traveling Salesman Problem (TSP) that we are going to explore now.

Note that the Travelling Salesman Problem is different from the Chinese postman problem as now the salesman must visit every city (vertex) rather than every street (edge). In fact this problem is still a challenge for mathematicians and so far there is no known algorithm that is simple enough to use to solve it. All we can do is list all possible routes and then decide which one is the shortest. This is called the **brute force** method but for a complete graph with n vertices we would need to analyze $\frac{(n-1)!}{2}$ cycles.

A Hamiltonian path (or traceable path) is a path in an undirected or directed graph that visits each vertex exactly once. A Hamiltonian cycle is a closed Hamiltonian path.

Not all graphs G have Hamiltonian cycles. In these cases we duplicate edges *to complete* the network, i.e. we may cross the same edge more than once to visit a vertex.

Example 7

Consider this Hamiltonian graph G.
List all possible Hamiltonian cycles and their total weight. Hence state the optimal travelling salesman solution for the graph G.

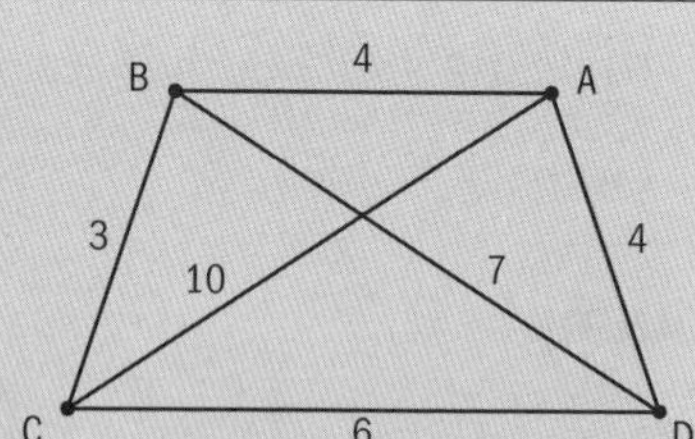

ABCDA (or ADCBA) has weight 17.
ACDBA (or ABDCA) has weight 27.
ADBCA (or ACBDA) has weight 24.

ABCDA or ADCBA or BADCB … have the same weight as they are just permutations of the same edges. It is enough to list all cycles starting with A and watch for reverse order of the edges.

The trouble with this brute force approach is that as the number of cities grows, the corresponding number of round-trips to check quickly outstrips the capabilities of the fastest computers. With 10 cities, there are more than 300 000 different round-trips. With 15 cities, the number of possibilities soars to more than 87 000 000 000.

So, this method is not suitable to tackle everyday problems if we want to solve them in a reasonable amount of time. The solution is to compromise and try to find lower and upper bounds for an optimal solution. The lower bound is the minimum distance that we must travel and the upper bound is the maximum. If we find a lower bound of 25 and an upper bound of 27, then the optimal answer must be between these two numbers. So we want to find a lower bound as *high* as possible and an upper bound as *low* as possible. If we are lucky and the bounds are the same number then this *must* be the answer to the problem!

The Nearest Neighbour Algorithm for upper bound

This is a basic, common-sense algorithm. You start at home and travel to the closest town that you have not yet visited. When you have visited every town you return home directly. It is the last part that can prove to be a long journey – the trip home. For this reason, this algorithm is not perfect but it is straightforward:

1. Start at a selected vertex (let's say home).
2. The next city will be the closest *as-yet-unvisited* one. (If there are two or more at the same closest distance, just pick any one of them).
3. Go there.
4. Repeat **2** and **3** until there are no unvisited cities.
5. Go back home.

Example 8

Ewa wants to travel from Vienna and visit all the cities represented on the graph here.

The labels of the edges represent the price, in euros, of the cheapest flight available between the cities listed. Find the upper bound for the total cost of her round trip.

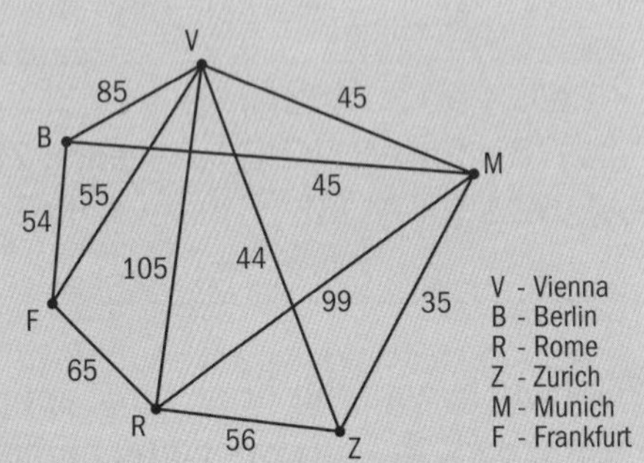

Start from Vienna (V). Go to Zurich (Z). Then to Munich (M), Berlin (B), Frankfurt (F), Rome (R) and return to Vienna (V). So the cycle is VZMBFV which corresponds to a cost of 44 + 35 + 45 + 54 + 65 + 105 = € 348.	*First vertex is V. VZ is the edge from V with minimum weight. From Z the minimum weight edge (apart from VZ) is MZ. From M the minimum weight edge to a city not yet visited is BM. From B the minimum weight edge to a city not yet visited is BF. Then the only vertex left is R.*

Deleted vertex algorithm for lower bound

1. Delete a vertex and find the minimum spanning tree for what remains.
2. Reconnect the vertex you deleted using the two edges with least weights.
3. Repeat this process for all vertices.
4. Select the *highest* total as the best lower bound.

Example 9

Ewa wants to travel from Vienna and visit all the cities represented on the graph.

The labels of the edges represent the price, in euros, of the cheapest flight available between the cities listed. Find the lower bound for the total cost of her round trip.

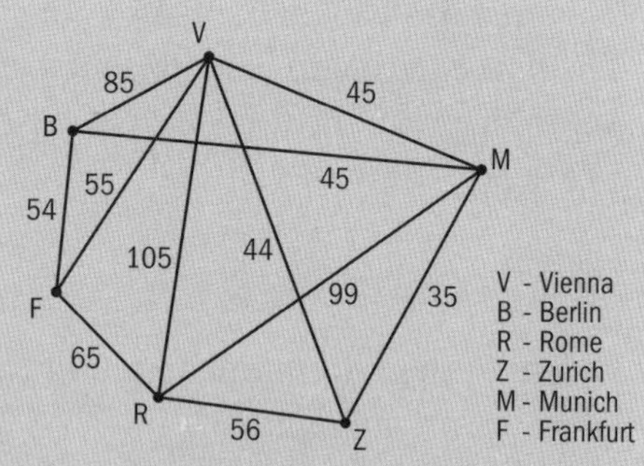

Apply the deleted vertex algorithm for lower bound to each vertex:

1 Delete a vertex and find the minimum spanning tree for what remains.

2 Reconnect the vertex you deleted using the two edges with least weights.

Delete V; the minimum spanning tree of $G - \{V\}$ is:

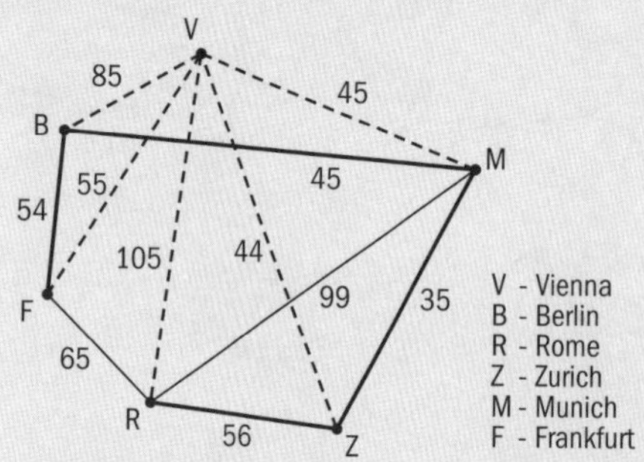

This tree has weight 35 + 45 + 54 + 56 = 190
The two edges with least weights are VZ and VM. So, in this case the total weight is 190 + 45 + 44 = 279

3 Repeat this process for all vertices.

Delete M; the minimum spanning tree of $G - \{M\}$ is shown on the right.
This tree has weight 44 + 54 + 55 + 56 = 209.
The two edges with least weight are MZ and MV. So, in this case the total weight is 209 + 35 + 45 = 289.

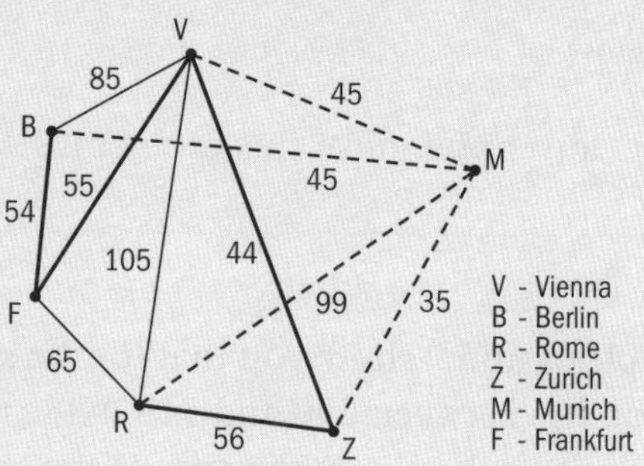

Delete Z; the minimum spanning tree of $G - \{Z\}$ is shown on the right.
This tree has weight 44 + 45 + 54 + 65 = 208.
The two edges with least weight are MZ and VZ. So, in this case the total weight is
208 + 35 + 44 = 287.

Delete R; the minimum spanning tree of $G - \{R\}$ is shown on the right.
This tree has weight 35 + 44 + 45 + 54 = 178.
The two edges with least weight are RZ and FR.
So, in this case the total weight is
178 + 56 + 65 = 299.

Delete F; the minimum spanning tree of $G - \{F\}$ is shown on the right.
This tree has weight 35 + 44 + 45 + 56 = 180.
The two edges with least weight are FB and FV.
So, in this case the total weight is
180 + 54 + 55 = 289.

Delete B; the minimum spanning tree of $G - \{B\}$ is shown on the right.
This tree has weight 35 + 44 + 55 + 56 = 190.
The two edges with least weight are BM and BF.
So, in this case the total weight is
190 + 45 + 54 = 289.

So the lower bound for this problem is 299.

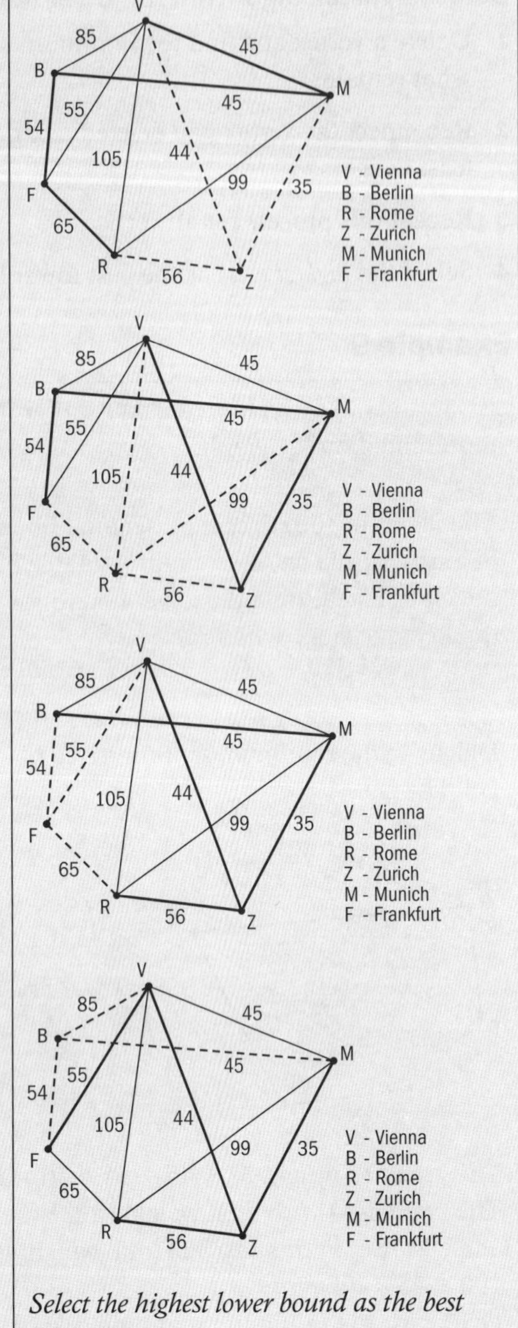

Select the highest lower bound as the best lower bound.

Examples 8 and 9 show you that for the situation given the optimal solution lies between 299 and 348. Note that in the context given the algorithms analyze only the cost of the trip. In real life many other factors are taken into account when selecting flights. In general, search engines allow users to set additional conditions like times of flights, duration, number of stopovers and order possibilities according to selected criteria. However, in most cases, due to limitations of the algorithms available, the decision needs to be made by the user based on a much smaller number of possibilities than the ones dealt with by the algorithms behind the machine.

The following example shows you a case where the lower bound and the upper bond are the same, allowing you to find an optimal solution for the Travelling Salesman Problem.

Example 10

A complete graph G of order 5 has edges with the weights shown in the diagram.

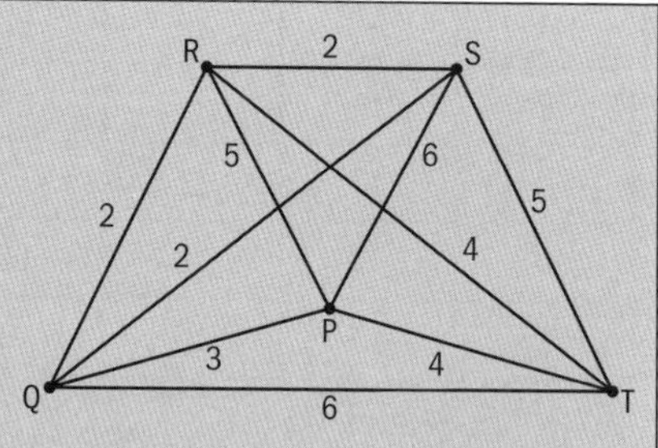

Consider the Travelling Salesman Problem for G.

a Explain why the total weight of the cycle PQSRTP is an upper bound for the Travelling Salesman Problem for G.

b By removing the vertex P, find a lower bound for the Travelling Salesman Problem for G.

c Hence state the total length of the solution to the salesman problem for G.

---

a The length of any Hamiltonian cycle is always an upper bound for the Travelling Salesman Problem of a graph.

The upper bound is any value greater than or equal to the minimum total weight of a Hamiltonian cycle of the graph.

Note that the cycle given could be found using the nearest neighbour algorithm starting from P and then chosing S (but not R).

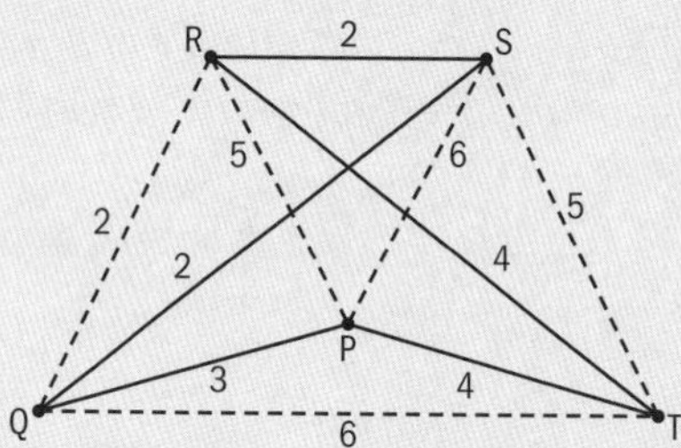

b Delete P; the minimum spanning tree for $G - \{P\}$ is shown on the right.
This tree has weight 2 + 2 + 4 =8.
The two edges with least weight are PQ and PT. So, a lower bound for the Travelling Salesman Problem of G is 8 + 3 + 4 = 15.

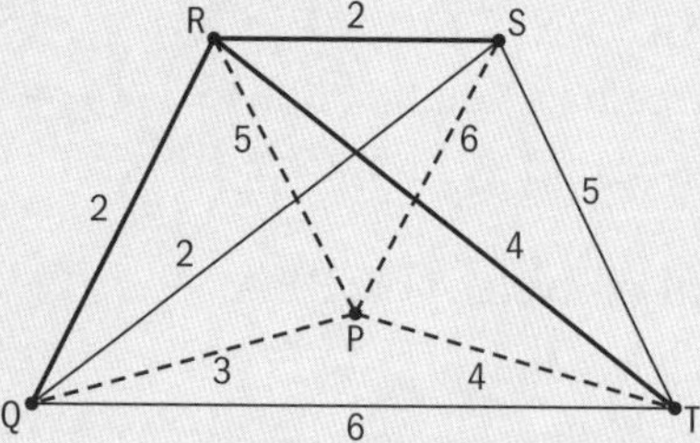

Add the total weight of the tree with the weights of the two edges with least weight.

c As the lower and upper bounds found in parts **a** and **b** are equal to 15, the solution to the Travelling Salesman Problem for this graph must have length 15.

Note that the cycle PQSRT is a solution of the Travelling Salesman Problem as its total weight is equal to the lower and upper bounds found.

Sometimes we come across a Hamiltonian cycle that has a total weight equal to a lower bound. This is another situation that allows us to solve the Travelling Salesman Problem for a graph easily, as shown in the following example.

Example 11

The complete graph *H* has this cost adjacency table.

	A	B	C	D	E
A	–	19	17	10	15
B	19	–	11	16	13
C	17	11	–	14	13
D	10	16	14	–	18
E	15	13	13	18	–

a By first finding a minimum spanning tree on the subgraph of *H* formed by deleting vertex A and all edges connected to A, find a lower bound for this problem.

b Find the total weight of the cycle ADCBEA.

c Hence state the optimal solution for the Travelling Salesman Problem. Explain.

a

	A	B	C	D	E
A	–	–	–	–	–
B	–	–	11	16	13
C	–	11	–	14	13
D	–	16	14	–	18
E	–	13	13	18	–

Apply the deleted vertex algorithm for lower bound to each vertex:

1 Delete a vertex and find the minimum spanning tree T for what remains.

$T = \{BC, BE, CD\}$
This tree has weight $11 + 13 + 14 = 38$.
Lower bound $= 38 + 10 + 15 = 63$.

2 Add the weight of the two edges with least weights that had been deleted from the first row/column of the original table.

b Weight of cycle ADCBEA
$= 10 + 14 + 11 + 13 + 15 = 63$.

Add the weights of all the edges in this cycle.

c ADCBEA gives a solution to the Travelling Salesman Problem for the graph *G* because its total weight is equal to the lower bound.

The cycle ADCBEA has optimal weight.

Exercise 5D

1 Consider the weighted graph here.

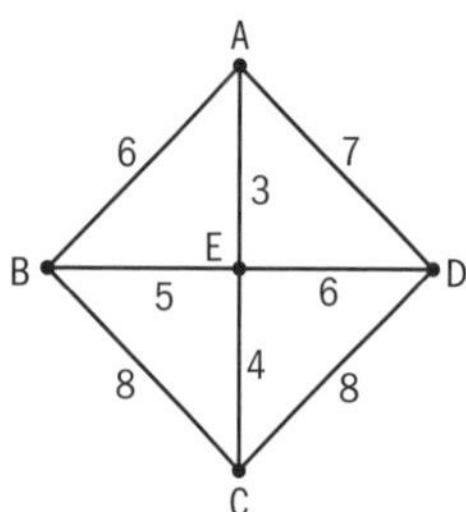

a Find a lower bound for the Travelling Salesman Problem for this graph.

b Find an upper bound for the Travelling Salesman Problem for this graph.

2 Consider the graph below where the vertices represent cities to be visited by a salesman. The weight of the each edge indicates the distance between the cities incident with the edge.

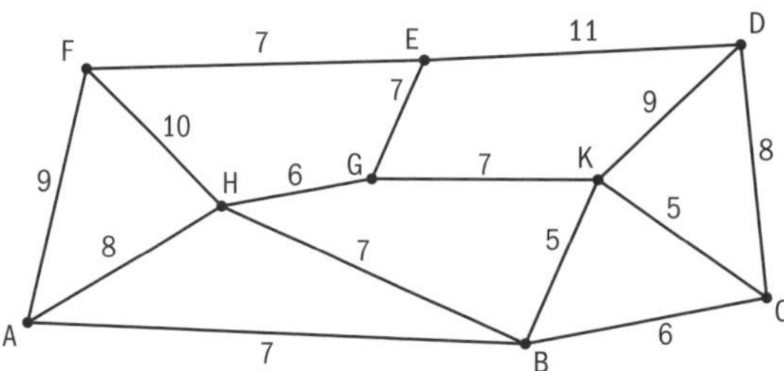

a Use the Nearest Neighbour Algorithm for determining a least upper bound for the Travelling Salesman Problem.

b Use the Deleted Vertex algorithm for determining a lower bound for Travelling Salesman Problem.

3 Let G be the graph to the right.

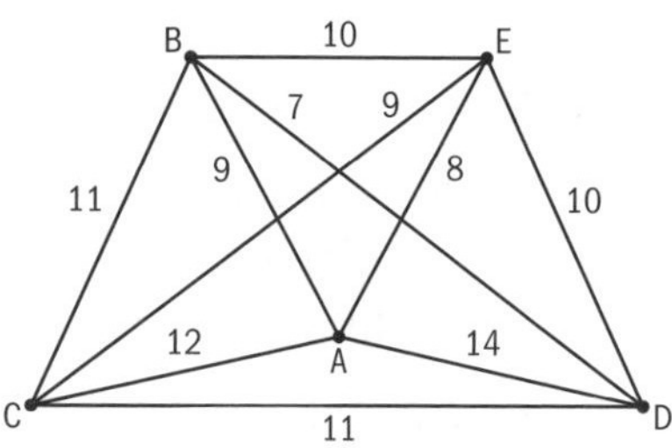

a State the order and size of the graph G. Hence, state whether or not the graph is complete.

b Find the total number of Hamiltonian cycles in G, starting at vertex A. Explain your answer.

c **i** Find a minimum spanning tree for the subgraph obtained by deleting A from G.

ii Hence, find a lower bound for the Travelling Salesman Problem for G.

d Give an upper bound for the Travelling Salesman Problem for the graph.

e Show that the lower bound you have obtained is not the best possible for the solution to the Travelling Salesman Problem for G.

Review exercise

EXAM-STYLE QUESTIONS

1 Graph G has vertices A, B, C, D and E. Its adjacency table is:

	A	B	C	D	E
A	0	2	1	1	0
B	2	2	1	1	0
C	1	1	2	0	2
D	1	1	0	0	0
E	0	0	2	0	0

a Draw the graph G.

b **i** Define an Eulerian circuit.

ii Write down an Eulerian circuit in G starting at A.

c **i** Define a Hamiltonian cycle.

ii Explain why it is not possible to have a Hamiltonian cycle in G.

2 The following diagram shows a weighted graph G.

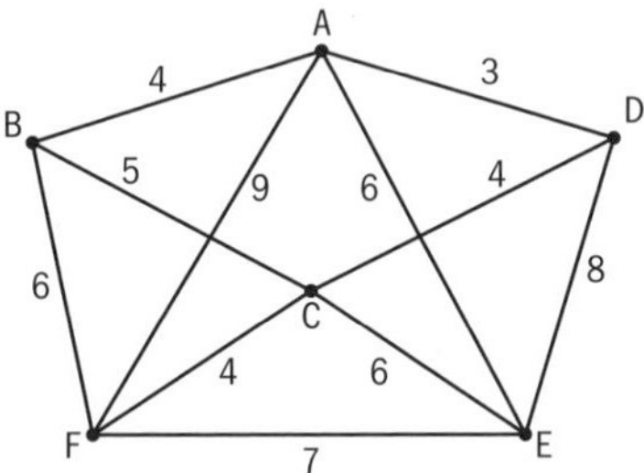

a **i** Explain briefly what features of the graph enable you to state that G has an Eulerian trail but does not have an Eulerian circuit.

ii Write down an Eulerian trail in G.

b **i** Use Kruskal's algorithm to find and draw the minimum spanning tree for G. Your solution should indicate the order in which the edges are added.

ii State the weight of the minimum spanning tree.

c Use Dijkstra's Algorithm to find the path of minimum total weight joining A to each other vertex in the graph, and show the weight of each path. Your solution should clearly indicate the use of this algorithm.

3 The graph G has the following cost adjacency table.

	A	B	C	D	E	F
A	–	20	–	–	–	89
B	20	–	65	–	–	31
C	–	65	–	73	31	20
D	–	–	73	–	12	–
E	–	–	31	12	–	73
F	89	31	20	–	73	–

a Draw G in planar form.

b Use Dijkstra's Algorithm to find the shortest path between the vertices A and D. Show all the steps in the algorithm and state the length of the shortest path.

4 The graph G has the cost adjacency table shown here.

	A	B	C	D	E
A	–	10	–	9	5
B	10	–	8	–	3
C	–	8	–	8	4
D	9	–	8	–	6
E	5	3	4	6	–

a Draw G in planar form.

b List all the distinct Hamiltonian cycles in G beginning and ending at A. (When one cycle is the reverse of another, you may consider the two identical.) Hence determine the Hamiltonian cycle of least weight.

c Giving a reason, determine the maximum number of edges that could be added to G while keeping the graph both simple and planar.

5 The weights of the edges of a graph with vertices P, Q, R, S and T are given in this cost adjacency table.

	P	Q	R	S	T
P	–	20	25	21	26
Q	20	–	22	29	23
R	25	22	–	28	24
S	21	29	28	–	27
T	26	23	24	27	–

a Find an upper bound for the Travelling Salesman Problem for this graph.

b **i** Use Kruskal's algorithm to find and draw a minimum spanning tree for the subgraph obtained by removing the vertex T from the graph.

ii State the total weight of this minimum spanning tree and hence find a lower bound for the Travelling Salesman Problem for this graph.

EXAM-STYLE QUESTION

6 The weighted graph G is shown here.

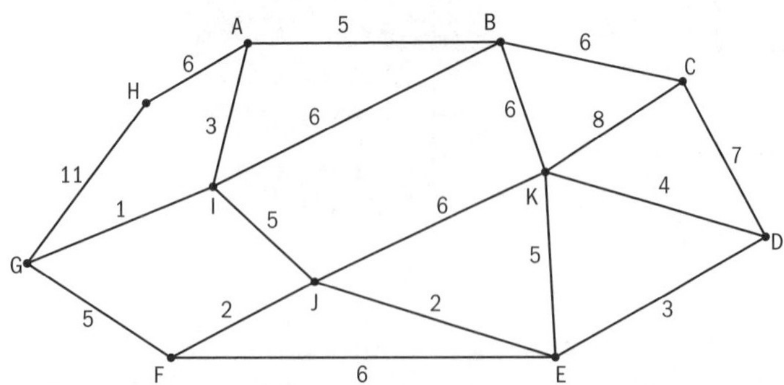

Consider the subgraph G' of G obtained by deleting vertex H from G.

a Use Kruskal's algorithm to find the minimum spanning tree of graph G' and state its weight.

b Hence find the weight of a lower bound for the Hamiltonian cycle in G beginning at vertex H.

c Prove that to find the Hamiltonian cycle of least weight for the complete graph K_n with $n > 3$, at most $\frac{1}{2}(n-1)!$ Hamiltonian cycles need to be examined.

d Hence state the number cycles in G would have to be examined to find the one with the least weight.

Chapter 5 summary

Kruskal's algorithm to find a **minimum spanning tree** T, of a weighted graph G with n vertices:

1. Select the edge with minimum weight to be the first edge of T.
2. Consider the weighted edges of $G - T$ which do not form a cycle with the already chosen edges of T. Pick the one with minimum weight and add the new edge and vertex to T (in case there is more than one with minimum weight, choose any of them).
3. Repeat step **2** until $n - 1$ edges have been chosen.

Dijkstra's Algorithm to find the **shortest path** between a selected root vertex and all other vertices: Start with an initial vertex. Let the distance of a vertex be the distance from the initial vertex to it.

1. Assign to every vertex a distance value: zero for the initial vertex and infinity for all other vertices.
2. Mark all vertices as unvisited. Set the initial vertex as current.
3. For the current vertex, consider all its unvisited neighbours and calculate the distance to each one of them. If this distance is less than the previously recorded distance then overwrite this distance.
4. After considering all neighbours of the current vertex, mark it as visited. A visited vertex will not be checked ever again; its recorded distance is now final and minimal.
5. Set the unvisited vertex with the smallest distance (from the initial vertex) as the next "current vertex" and continue from step **3**.
6. When all the vertices have been visited, STOP.

The Chinese Postman problem requires the shortest tour of a graph which visits each edge at least once. For an Eulerian graph, an Eulerian cycle is the optimal solution. The Traveling Salesman Problem requires the least total weight Hamiltonian cycle a salesman can take through each of n given cities. No efficient general method for finding the solution is known yet.

Chinese Postman algorithm to find the shortest Eulerian cycle

1 List all odd vertices.

2 List all possible pairings of odd vertices.

3 For each pairing, find the edges that connect the vertices with the minimum weight.

4 Find the pairings such that the sum of the weights is minimized.

5 On the original graph, add the edges that have been found in Step **4**.

6 The length of an optimal Chinese postman route is the sum of all weights of the edges including the ones added to the total in step **4**.

7 Draw the route corresponding to this minimum weight.

The Travelling Salesman problem

The Nearest Neighbour Algorithm for upper bound

1 Start at a selected vertex (let's say home)
2 The next city will be the closest *as-yet-unvisited* one. (If there are two or more at the same closest distance, just pick any one of them).
3 Go there.
4 Repeat **2** and **3** until there are no more unvisited cities.
5 Go back home.

Deleted vertex algorithm for lower bound

1 Delete a vertex and find the minimum spanning tree for what remains.
2 Reconnect the vertex you deleted using the two edges with least weights.
3 Repeat this process for all vertices.
4 Select the *highest* total as the best lower bound.

Answers

Chapter 1

Exercise 1A

1 **a** 3392 **b** 964596 **c** 3489
2 **a** 14416_{16} **b** 4321120_5 **c** 101011011_2
3 **a** 155_6 **b** 22201_6 **c** 1005_6
4 **a** 3052_7 **b** BDD_{16} **c** 10010001_2
5 **a** 221212_5 **b** 101010_2 **c** 3140_9
6 $b = 7$
7 **a** $n = 5$ **b** $(23_8)^2 = 551_8$

Exercise 1B

1 **a** 6 **b** 4 **c** 24 **d** 3
2 **a** $m = 28, n = -103$ **b** $m = 67, n = -89$
c $m = 5, n = -13$ **d** $m = 93, n = -238$

Investigation: Diophantus Riddle

3 $n = 84$ years old

Exercise 1C

1 **i** $x_0 = -1, y_0 = 2$ **ii** $x = -1 + 3k, y = 2 - 5k$
2 No solution
3 **i** $x_0 = 45, y_0 = -15$ **ii** $x = 45 + 8k, y = -15 - 3k$
4 No solution
5 **i** $x_0 = -4, y_0 = 4$ **ii** $x = -4 + 3k, y = 4 - 2k$
6 No solution
7 **i** $x_0 = 0, y_0 = 9$ **ii** $x = k, y = 9 - 6k$
8 **i** $x_0 = 40, y_0 = -80$
ii $x = 40 + 12k, y = -80 - 25k$
9 **i** $x_0 = -201000, y_0 = 1494000$
ii $x = -201000 + 238k, y = 1494000 - 1769k$
10 **i** $x_0 = -521534, y_0 = 1060041$
ii $x = -521534 + 1137k, y = 1060041 - 2311k$
11 No solutions for $c = \{11, 13, 14, 15, 16, 17, 19\}$
When $c = 12$, the general solution is:
$x = 118 + 165k, y = -10 - 14k$.
When $c = 18$, the general solution is:.
$x = -153 + 165k, y = 13 - 14k$.

12 Gino can buy either 4 bags of dog treats and 3 bags of cat treats or 1 bag of dog treats and 7 bags of cat treats.

Exercise 1D

1 493 is not prime; it has 17 and 29 as factors.
2 $19152 = 2^4 \times 3^2 \times 7 \times 19$
3 $m = 4, n = -9$

Exercise 1F

1 **b** $75420 = 2^3 \times 3^2 \times 5 \times 11 \times 19$
4 $a = 420$
5 **a** $n = 2 \times 3^2 \times 5^5 \times 7 = 3150$ **b** 210
6 **b** $\gcd(p, q) = 2^2 \times 3 \times 5 \times 7 \times 11 \times 13^2$

Review Exercise

1 **a** $m = 9A0, n = 650$
2 **b** 17, 19, 23, 29, 37, 47, 59, 73, 89, 107
c Counter example: $n = 17 \Rightarrow n^2 + n + 17 = 17 \times 19$
7 $x = 27 + 79k, y = 122 + 357k$
8 **a** $(A, B) = \{ (9, 71), (30, 40), (51, 9)\}$ **b** (9, 71)

Chapter 2

Skills check

2 **a** TBC **b** TBC **3** TBC

Exercise 2A

1 **a** 1 **b** 2 **c** 16
2 **a** No **b** No **c** No **d** Yes
e No
8 **a** Yes **b** 8
11 **c** Converse is not true

Exercise 2B

1 **a** 17 **b** No inverse **c** No Inverse **d** 3
2 **a** 7 **b** 8 **c** 11
3 **a** $28k + 24$ **b** $71k + 16$ **c** $133k + 101$
4 **a** $k = 4$
5 **a** $d = 4$ **b** $A = 66, B = -71$ **c** $x = 156$

Investigation - Mersenne primes and perfect numbers

n	$\sum_{k=0}^{n} 2^k$	$2^{n+1} - 1$	Prime	$2^n \times \sum_{k=0}^{n} 2^k$	Perfect Number
1	$1 + 2 = 3$	$2^2 - 1 = 3$	YES	$2 \times 3 = 6$	$1 + 2 + 3 = 6$
2	$1 + 2 + 4 = 7$	$2^3 - 1 = 7$	YES	$4 \times 7 = 28$	$1 + 2 + 4 + 14 + 7 = 28$
3	$1 + 2 + 4 + 8 = 15$	$2^4 - 1 = 15$	NO	$8 \times 15 = 120$	No
4	$1 + 2 + 4 + 8 + 16 = 31$	$2^5 - 1 = 31$	YES	$16 \times 31 = 496$	$1 + 2 + 4 + 8 + 16 + 31 + 62 + 124 + 248 = 496$
5	$1 + 2 + 4 + 8 + 16 + 32 = 63$	$2^6 - 1 = 63$	No	$32 \times 63 = 2016$	No

Conjecture $\sum_{k=0}^{n} 2^k = 2^{n+1} - 1$

If $2^n - 1$ is a prime number, then $2^{n-1}(2^n - 1)$ is a perfect number.

Exercise 2C

1 **a** 2 **b** 13 **c** 2601

Exercise 2D

1 $x = 2 + 143k$ **2** $x = 73 + 84k$
3 $x = 4 + 15k$
4 **a** $x = 51 + 140k$ **b** $x = 58 + 60k$
c $x = 45 + 154k$
5 8 **6** 11 **7** 316

Exercise 2E

1 **a** 3 **b** 26 **c** 31
2 12
3 **a** 1 **b** 10
4 **a** $x = 3 + 13k$ **b** $x = 6 + 7k$ **c** $x = 9 + 11k$
5 8 **8** **b** 82

Review Exercise

1 **a** $x = 1 + 10k$ **b** $x = 6 + 15k$
2 **c** 4, 7
3 **a** $m = 194, n = -25$ **b** 194
c $x = 776 + 1001k$ **d** no solution
4 **a** $x = 651 + 715k$, for $k \in \mathbb{Z}$.
b 1366, 2081, 2796
5 **a** $n = 1390_{11}$, $m = 2270_{11}$
7 **a** $109368 = 2^3 \times 3^2 \times 7^2 \times 31$
8 **c** 1 **9** $x = 5, y = 4$

Chapter 3

Skills check

1 **a** 2, 3, 2 and 3 **b** 4, 6, 10 and 18
2 **a** 1 and −2 **b** 0 and 1
3 18 and 3 **4** $\frac{5}{6}$

Investigation on Fibonacci numbers

a The quotient between two consecutive terms converges quickly to Φ. This can be observed with the help of a GDC spreadsheet. For example:

	A n	B fn	C qn	D
◆				
19	18	4181	1.6180338	
20	19	6765	1.6180341	
21	20	10946	1.618034	
22	21	17711	1.618034	
23	22	28657	1.618034	

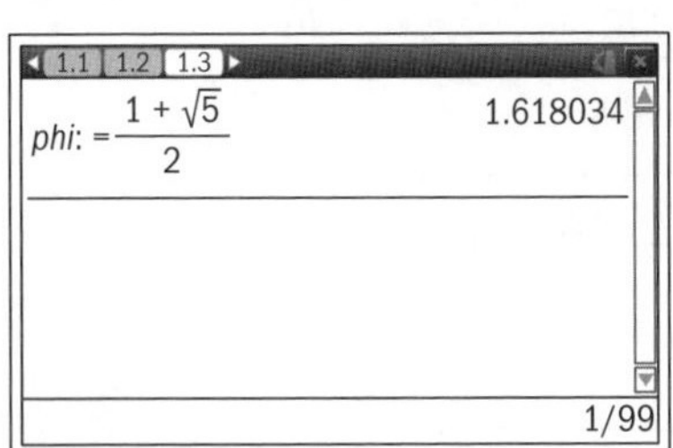

b Repeat part **a** using other initial values for the sequence. For example:

		B fn	C qn	D
◆				
1	0	5	_	
2	1	3	0.6	
3	2	8	0.6	
4	3	11	2.6666667	
5	4	19	1.375	

		B fn	C qn	D
◆				
19	18	15737	1.6180336	
20	19	25463	1.6180341	
21	20	41200	1.6180339	
22	21	66663	1.618034	
23	22	107863	1.618034	

e Using results from parts **c** and **d** we have
$1 + \Phi = \Phi^2$, $1 + \Phi^2 = \Phi^3$,
$1 + 3\Phi = (1 + 2\Phi) + \Phi = \Phi^3 + \Phi = \Phi(\Phi^2 + 1)$
$= \Phi\Phi^3 = \Phi^4, \ldots$
Therefore can re-write the sequence 1, Φ, 1 + Φ, 1 + 2Φ, 2 + 3Φ, 3 + 5Φ,… as 1, Φ, Φ^2, Φ^3, Φ^4, … which is a geometric sequence with first term 1 and common ratio Φ.

f Let $\Phi' = -\frac{1}{\Phi} = -\frac{2}{1+\sqrt{5}} = -\frac{2(1-\sqrt{5})}{(1+\sqrt{5})(1-\sqrt{5})} = \frac{1-\sqrt{5}}{2}$,
i.e. Φ′ is the negative of the reciprocal of the Golden Ratio Φ.

As $x^2 - x - 1 = 0 \Rightarrow \frac{1 \pm \sqrt{5}}{2}$. Therefore Φ′ is the other solution of $x^2 - x - 1 = 0$ and using the results in (c) – (e) we can conclude that $1 + \Phi' = \Phi'^2$, $1 + \Phi'^2 = \Phi'^3$, … and the reciprocal Golden sequence can be written as
1, Φ′, $(\Phi')^2$, $(\Phi')^3$, $(\Phi')^4$,…

Exercise 3A

2 **a** $a_1 = 2$, $a_2 = 2^2 = 4$, $a_3 = 4^2 = 16$, $a_4 = 16^2 = 256$ and $a_5 = 256^2 = 65536$.
b $a_n = 2^{2^{n-1}}, n \in \mathbb{Z}^+$
5 **a** 1, 2, 3, 5, 8, 13, 21
b

n	$\sum_{i=0}^{n} F_i^2$	$F_n \times F_{n+1}$
1	$0^2 + 1^2 = 1$	$1 \times 1 = 1$
2	$0^2 + 1^2 + 1^2 = 2$	$1 \times 2 = 2$
3	$0^2 + 1^2 + 1^2 + 2^2 = 6$	$2 \times 3 = 6$
4	$0^2 + 1^2 + 1^2 + 2^2 + 3^2 = 15$	$3 \times 5 = 15$
5	$0^2 + 1^2 + 1^2 + 2^2 + 3^2 + 5^2 = 40$	$5 \times 8 = 40$

Exercise 3B

1 a $u_n = 3^{n-1}\cdot 2 - 2\cdot\frac{3^{n-1}-1}{3-1} \Rightarrow u_n = 3^{n-1}+1$

b $u_n = 2^{n-1}\cdot 3 - 1\cdot\frac{2^{n-1}-1}{2-1} \Rightarrow u_n = 2^n+1$

c $u_n = (-1)^n\cdot 2 + 2\cdot\frac{(-1)^n-1}{-1-1} \Rightarrow u_n = (-1)^n+1$

d $u_n = (-2)^n\cdot(-1) + 2\cdot\frac{(-2)^n-1}{-2-1} \Rightarrow u_n = (-2)^{n+1}+1$

2 a $u_n = 11\cdot 2^{n-1} - 3n - 7$ **b** $u_n = \frac{5}{4}\cdot 3^n - \frac{1}{2}n - \frac{5}{4}$

3 b $u_n = n^2+1$

c $u_{n-1} = (n-1)^2+1$ or $u_{n-1} = n^2-2n+2$

4 b $u_n = n^3 - \frac{3}{2}n^2 + \frac{1}{2}n + 10$

5 $v_n = n$ and $u_n = \sqrt{n}$

7 $u_n = 17\cdot 2^n - 3n^2 - 12n - 18$ **8** $u_n = \frac{5^n - 3^{n+1}}{2}$

Exercise 3C

1 a Option B is better if $L \geq 24370.45$ (2 d.p.)

b 629407.91 (2 d.p.)

2 € 2074.35 **3** $L = 41135.72$ (2 d.p.)

4 a $r = 1.002263\ldots$ (Approximately 0.226% per month).

b $A = 521.75$ (2 d.p.) **c** 1825.80

5 a $w_1 = \frac{4}{7}$ and $w_2 = \frac{25}{49}$ **b** $w_n = \frac{3}{7} + \frac{1}{7}w_{n-1}$

c $w_n = \frac{1}{14}\left(\frac{1}{7}\right)^{n-1} + \frac{1}{2}$ **d** $\frac{1}{2}$

Exercise 3D

1 a $u_n = \frac{2\cdot 4^n + 3\cdot(-1)^n}{5}$ **b** $u_n = \frac{5+(-1)^n}{2}$

c $u_n = (n+2)\cdot 2^{n-1}$

d $u_n = \cos(n\theta) + \frac{\cos\theta - 1}{\sin\theta}\cdot\sin(n\theta)$ where $\theta = \arctan 2$.

2 a i $u_{n+2} - u_{n+1} = 2(u_{n+1} - u_n)$ **ii** $u_n = 3\cdot 2^n$

b i $u_{n+2} - u_{n+1} = 4(u_{n+1} - u_n)$ **ii** $u_n = \frac{2+4^n}{3}$

3 $P_n = \frac{20 - 8\cdot\left(-\frac{1}{2}\right)^n}{3}$

4 a There is just one way to get to step 1: therefore $V_1 = 1$.
There are two ways of getting to step 2: climb 1 step + 1 step or climbing two steps at a time. therefore $V_2 = 2$.
From the third step on, to obtain the number of way to get to the nth step, we just need to add the number of ways of getting to the $(n-2)$th step (and then go up two steps) and with the number of ways of getting to the $(n-1)$th step (and then go up one step).
As these ways of getting to the nth step are mutually exclusive, $V_n = V_{n-1} + V_{n-2}$, for $n \geq 3$.

b $V_n = F_{n+1}$, for $n \geq 1$ **c** $V_n = \frac{\Phi^{n+1} - (\Phi')^{n+1}}{\sqrt{5}}$

6 $L_n = \Phi^n + (\Phi')^n$

Review Exercise

1 a $u_n = (-3)^n + \frac{(-3)^n - 1}{-3-1} \Rightarrow u_n = \frac{3}{4}(-3)^n + \frac{1}{4}$.

b $u_n = 6\cdot 5^n - 2\frac{5^n-1}{5-1} \Rightarrow u_n = \frac{11\cdot 5^n + 1}{2}$.

c $u_n = 9\cdot 6^n + n + 1$ **d** $u_n = \frac{-41\cdot 7^n + 12n + 14}{9}$

4 $u_n + A\cdot u_{n-1} + B\cdot u_{n-2} = C,\ n \geq 2$

a $A = -6$, $B = 7$ and $C = -7$

b $u_5 = 2$, $u_6 = -79$ and $u_7 = -495$

5 $u_n = \frac{2}{3}3^n\left(\sin\left(\frac{n\pi}{2}\right) - 6\cos\left(\frac{n\pi}{2}\right)\right)$

6 a $a_n = 1.08a_{n-1}$ and $a_0 = 1000$

b $b_n = 100 + 1\cdot 08b_{n-1}$ and $b_0 = 100$

c $a_n = (1.08)^n\cdot 1000$ and $b_n = (1.08)^n\cdot 1350 - 1250$

d at least 17 years.

7 $u_n = \left(1+\sqrt{3}+(-1)^n\left(1-\sqrt{3}\right)\right)\left(\sqrt{3}\right)^n$

8 $v_n = \frac{(-1)^{n+1} + 2^{n+2}}{3}$

9 The general solution is the form $u_n = A\cdot 2^n + B\cdot 3^n$.

$$u_n = 3^n \Rightarrow \begin{cases} A=0 \\ B=1 \end{cases} \Rightarrow \begin{cases} u_1 = 3 \\ u_2 = 9 \end{cases}$$

10 b $u_n = \left(4\cos(n\theta) + \frac{3-4\cos\theta}{\sin\theta}\cdot\sin(n\theta)\right)\cdot 4^n$

$v_n = (4\cos(n\theta) - 4\cos((n-1)\theta) + \frac{3-4\cos\theta}{\sin\theta}\cdot(\sin(n\theta) - \sin((n-1)\theta)))\cdot 4^n$

Chapter 4

Skills check

1 a $A \cap B = \{6\}$

b $A \cup B = \{2, 4, 5, 6, 7\}$

c $A' = \{1, 3, 5, 7, 8\}$

d $(A \cap B)' = \{1, 2, 3, 4, 5, 7, 8\}$

2

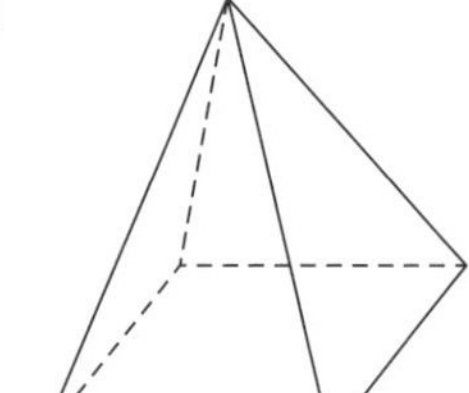

5 vertices, 8 edges and 5 faces.

3 At least 5 samples

Short Investigation on Handshaking Lemma

1 For example:

Graph	Order	Size	Degree of vertices
	4	4	In alphabetic order: 1, 3, 2, 2
	4	6 *(including the loop)*	In alphabetic order: 4, 3, 2, 3
	5	7	In alphabetic order: 3, 4, 2, 3, 2
…	…	…	…

2 The size of the graph is equal to half of the sum of the degrees of all the vertices.

3 and 4 Either all the vertices have even degree or the number of vertices with odd degree appear in pairs.

Exercise 4A

1 **a** 8 vertices; **b** 14 edges;

c *deg* (A) = 3, *deg* (B) = 3, *deg* (C) = 3, *deg* (D) = 6, *deg* (E) = 3, *deg* (F) = 3, *deg* (G) = 4 and *deg* (H) = 3.

d

	A	B	C	D	E	F	G	H
A	0	1	0	0	0	0	1	1
B	1	0	1	1	0	0	0	0
C	0	1	0	1	1	0	0	0
D	0	1	1	0	1	1	1	1
E	0	0	1	1	0	1	0	0
F	0	0	0	1	1	0	1	0
G	1	0	0	1	0	1	0	1
H	1	0	0	1	0	0	1	0

2 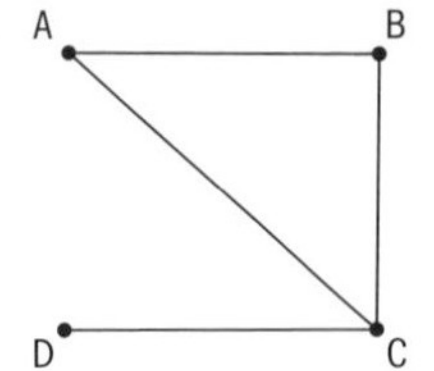

and

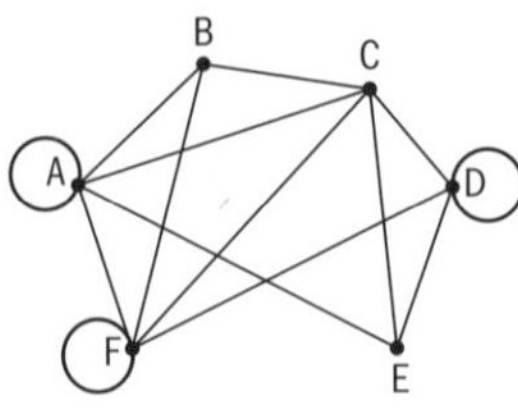

a the graph has size 4 and order 4.

b the graph has size 14 and order 6.

3 There are several possible answers to this question.

a Graph G_1 has 6 vertices and 10 edges and Graph G_2 has 6 vertices and 12 edges, for example:

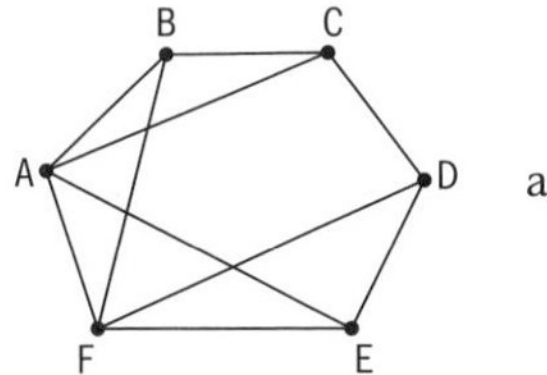

and

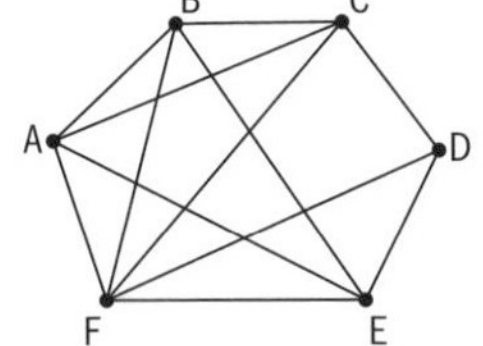

b The adjacency tables for these graphs are

G_1	A	B	C	D	E	F
A	0	1	1	0	1	1
B	1	0	1	0	0	1
C	1	1	0	1	0	0
D	0	0	1	0	1	1
E	1	0	0	1	0	1
F	1	1	0	1	1	0

G_2	A	B	C	D	E	F
A	0	1	1	0	1	1
B	1	0	1	0	1	1
C	1	1	0	1	0	1
D	0	0	1	0	1	1
E	1	1	0	1	0	1
F	1	1	1	1	1	0

4 12 vertices with degree 3 and 10 the number of vertices with degree 5.

Short Investigation on Trees

There are several possible trees with 5 vertices, e.g.

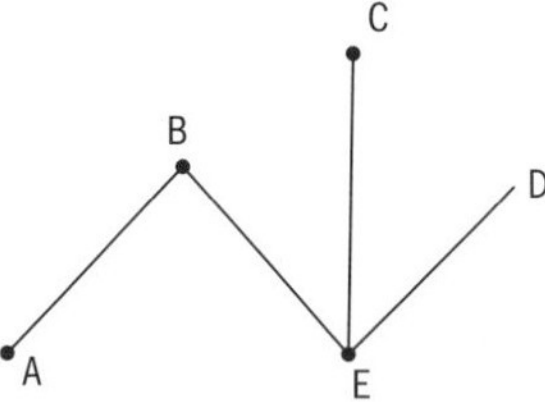

No matter which pair of vertices we choose, there is always only one path connecting them. The tree has 4 edges. If we remove one edge the graph becomes disconnected, e.g.

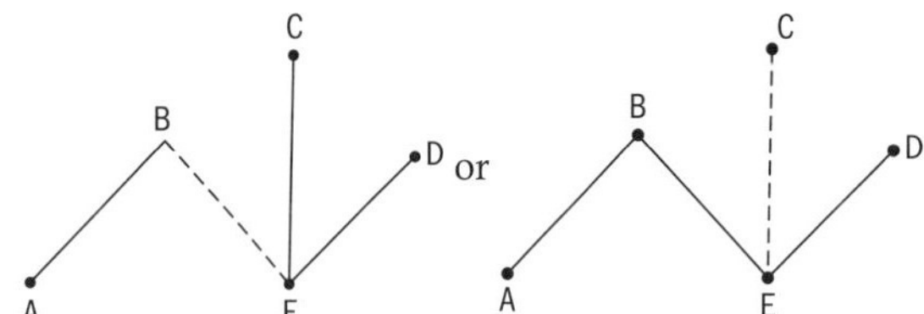

For any trees you draw with 6, 7 or 8 vertices, the number of edges is 5, 6 and 7 respectively. If you remove any edge from these trees you obtain a disconnected graph.

Exercise 4B

1 **a** 3 and 4 **b** 6 and 8 **c** 4 **d** 2

2

Graph	Order	Size
1	5	5
2	6	6
3	6	9
4	5	4
5	7	12
6	5	10
7	6	8
8	6	15

5 $d \in \{2, 4, 6, 8\}$

6 **a** $K_{3,4}$ has 7 vertices and 12 edges.

b $K_{13,17}$ has 30 vertices and 221 edges.

c $K_{12,5}$ has 17 vertices and 60 edges.

7 The sets of the partition have 8 and 16 vertices.

Exercise 4C

1 **a** Not isomorphic because they have different degree sequences.

b Not isomorphic because the first graph has three vertices of degree 2 and the second one has just two.

c Not isomorphic because they do not have the same number of edges.

2 Let the vertices be A, B and C. There is one possible simple graph with size 2 and four non-isomorphic multigraphs:

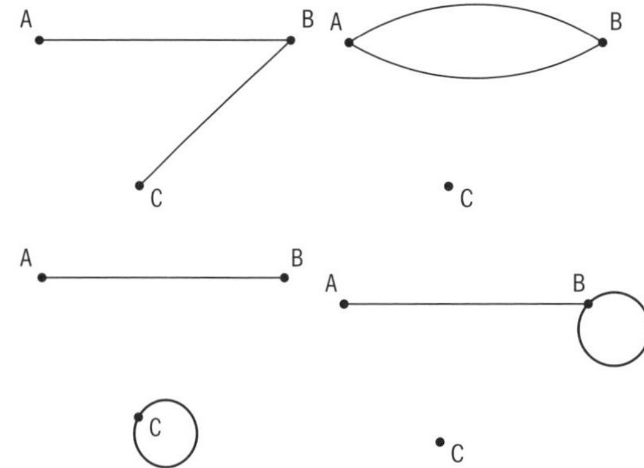

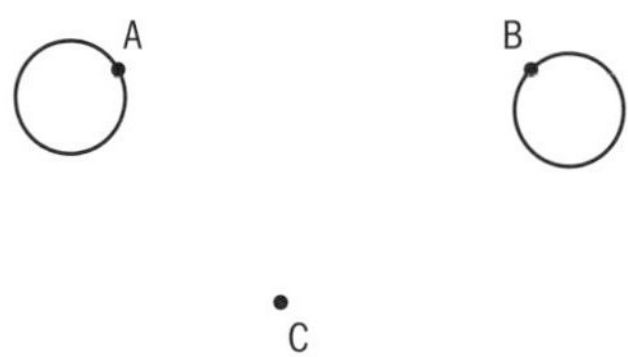

3 Let the vertices be A, B, C and D. There are three non-isomorphic simple graphs with size 3 and nine non-isomorphic multigraphs:

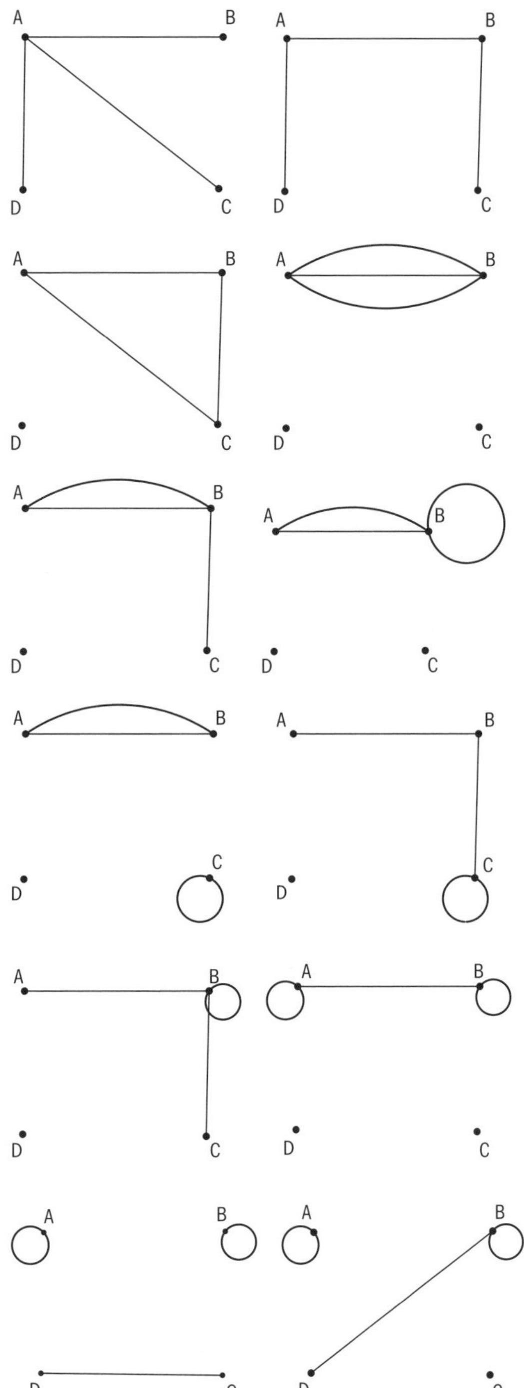

4

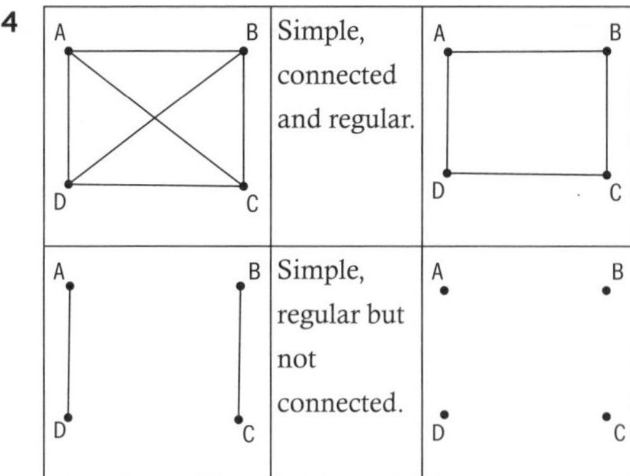

Exercise 4D

1

	Graph	Spanning tree	Another spanning tree
(a)			
(b)			
(c)			
(d)			

4 a It is possible to have a self-complementary graph of order 4 as shown in the diagram:

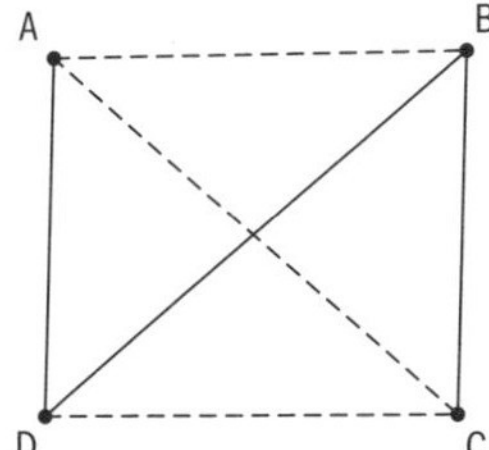

An isomorphism between the graph shown above and its complementary shown using dashed arcs is defined by A → B; B → C; C → D; and D → A

Mini Investigation on Euler relation for planar graphs

Polyhedron	Number of faces (f)	Number of edges (e)	Number of vertices (v)	$f + v$
Cube	6	12	8	14
Pentagonal prism	7	15	10	17
Tetrahedron	4	6	4	8
Hexagonal pyramid	7	12	7	14

Conjecture: $e + 2 = f + v$ (or equivalent).

Exercise 4E

1 a Graph G: $v = 6$ and $e = 9$; graph H: $v = 9$ and $e = 12$.

b Graph G: $6 - 9 + f = 2 \Rightarrow f = 5$; graph H: $9 - 12 + f = 2 \Rightarrow f = 5$.

2 $v = 8$, $f = 10$ and $e = 16$;

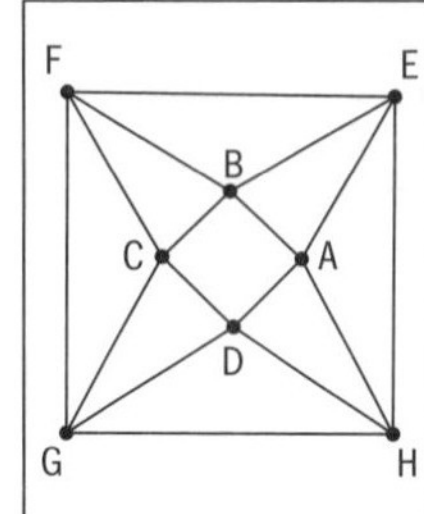

We can re-label the vertices in cyclic order and obtain a graph H isomorphic to G; the isomorphism that maps G onto H is defined by: A → A; B → B; C → E; D → H; E → D; F → C; G → F and H → G.

3 b The order of G is 12 and its size 30.

c 20 **d** G represents an icosahedron.

4 a i G_1 is bipartite: the set of its vertices can be split into two disjoint sets $M = \{A, C, E\}$ and $N = \{B, D, F\}$ such that each edge connects a vertex in M with the vertex in N.

ii G_1 and G_2 cannot be isomorphic as G_1 has no triangular cycles and G_2 does.

iii G_3 cannot be planar as it has 6 vertices, 9 edges and 6 quadrangular faces:

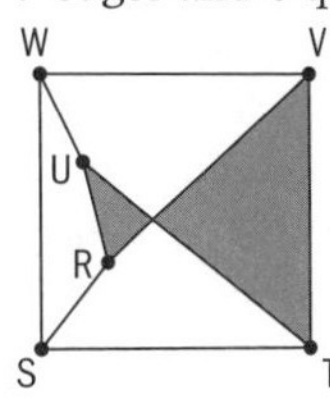

Face VTUR

Faces SRUT, TUWV and SRUW

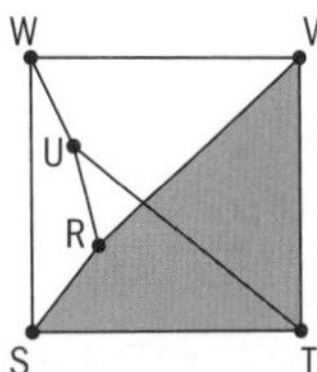

Faces RSTV and STVW

Therefore it does not fulfil the Euler identity for planar graphs.

You can also show that G_3 is isomorphic to G_1 and therefore non-planar.

b The complement of G_1 (which is isomorphic to $K_{3,3}$) has size 6.

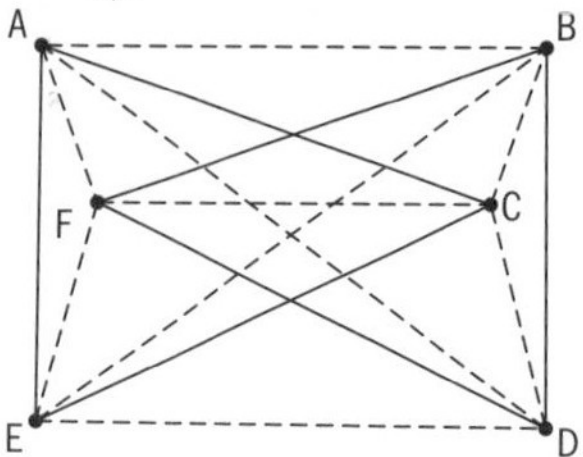

c G_2 is clearly a planar graph and therefore it can be represented by a polyhedron whose faces correspond to the regions bounded by the cycles of G_2:

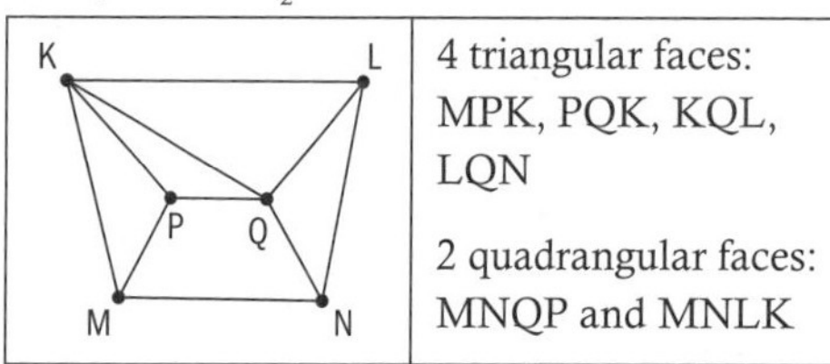

4 triangular faces: MPK, PQK, KQL, LQN

2 quadrangular faces: MNQP and MNLK

5 a 30 **b** 20 **c**

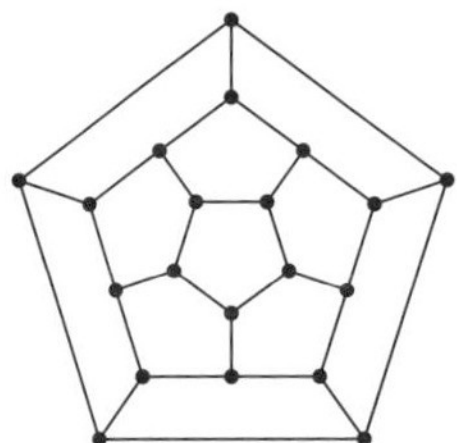

Exercise 4F

1 **a, b, c, d and g** are Hamiltonian graphs; **e** is semi- Hamiltonian graphs; **f** and **h** do not have any Hamiltonian paths.

2 There are 5! = 120 Hamiltonian cycles. Hint: Use software like Wolfram Alpha to generate them!

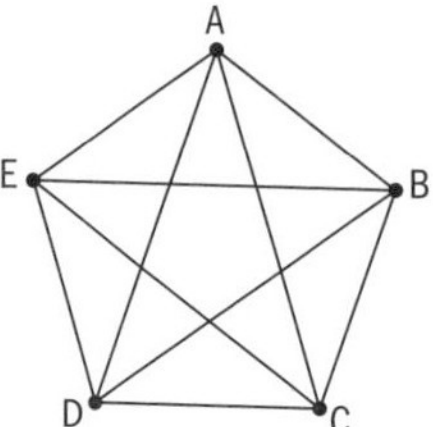

Exercise 4G

1 **a** is an Eulerian graph; **d**, **f** and **h** are semi-Eulerian.

2 a 12; **b** 12

3 a 18 **b** 36

4 a

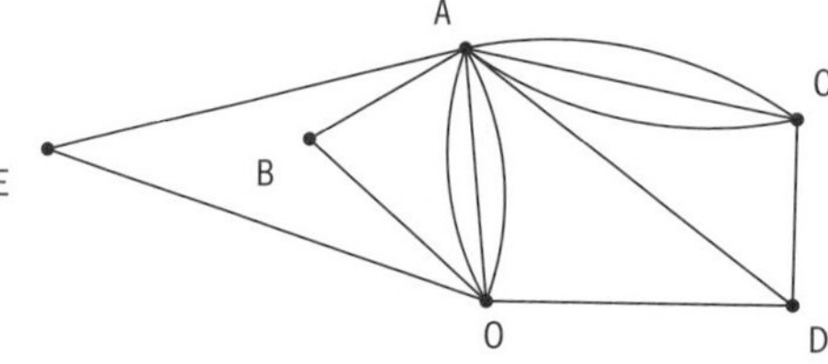

b No. **c** No.

5 c G is Hamiltonian. H does not have any Hamiltonian path.

6 b For order 4 it is possible to have both a graph and its complement semi-Eulerian as shown in the diagram below where the continuous lines represent a graph G and the dashed line its complement. These graphs are isomorphic and both are semi-Eulerian.

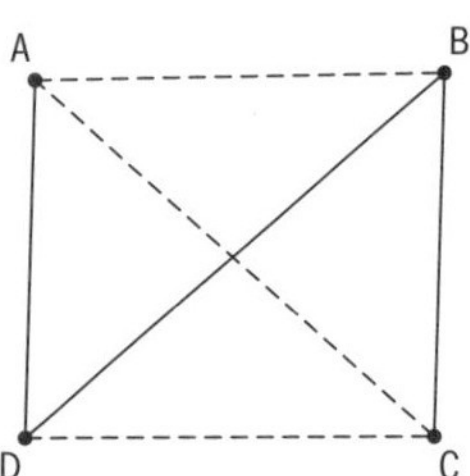

7 a

tetrahedron

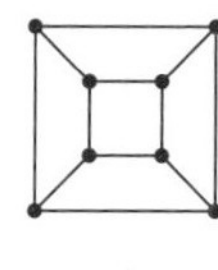
cube

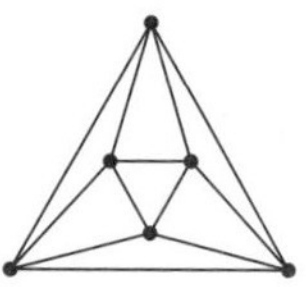
octahedron

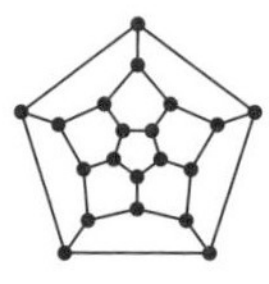
dodecahedron

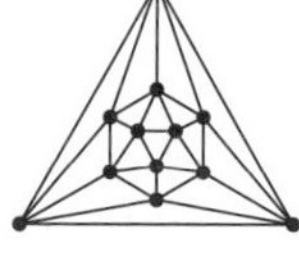
icosahedron

Review Exercise

1 a No **b** Yes

2

<table>
<tr><th>Graph</th><th>v</th><th>e</th><th>Degree sequence</th><th>Adjacency matrix</th></tr>
<tr><td>(1)</td><td>4</td><td>6</td><td>3, 3, 3, 3</td><td><table>
<tr><td></td><td>A</td><td>B</td><td>C</td><td>D</td></tr>
<tr><td>A</td><td>0</td><td>1</td><td>1</td><td>1</td></tr>
<tr><td>B</td><td>1</td><td>0</td><td>1</td><td>1</td></tr>
<tr><td>C</td><td>1</td><td>1</td><td>0</td><td>0</td></tr>
<tr><td>D</td><td>1</td><td>1</td><td>1</td><td>0</td></tr>
</table></td></tr>
<tr><td>(2)</td><td>4</td><td>11</td><td>5, 6, 5, 6</td><td><table>
<tr><td></td><td>A</td><td>B</td><td>C</td><td>D</td></tr>
<tr><td>A</td><td>2</td><td>1</td><td>1</td><td>1</td></tr>
<tr><td>B</td><td>1</td><td>2</td><td>1</td><td>2</td></tr>
<tr><td>C</td><td>1</td><td>1</td><td>0</td><td>3</td></tr>
<tr><td>D</td><td>1</td><td>2</td><td>3</td><td>0</td></tr>
</table></td></tr>
<tr><td>(3)</td><td>5</td><td>5</td><td>2, 2, 2, 1, 3</td><td><table>
<tr><td></td><td>A</td><td>B</td><td>C</td><td>D</td><td>E</td></tr>
<tr><td>A</td><td>2</td><td>0</td><td>0</td><td>0</td><td>0</td></tr>
<tr><td>B</td><td>0</td><td>0</td><td>1</td><td>0</td><td>1</td></tr>
<tr><td>C</td><td>0</td><td>1</td><td>0</td><td>0</td><td>1</td></tr>
<tr><td>D</td><td>0</td><td>0</td><td>0</td><td>0</td><td>1</td></tr>
<tr><td>E</td><td>0</td><td>1</td><td>1</td><td>1</td><td>0</td></tr>
</table></td></tr>
</table>

3 a 8 **b** No **c** $r < v$

4 All the cycles in bipartite graphs have even length due to the fact that each vertex from a set of the partition can just be adjacent to a vertex from the other set of the partition. So as we move around the graph, we can just return to the same vertex after travelling an even number of edges.

6 The graphs are not isomorphic because they have different degree sequences: one graph has two adjacent vertices with degree 2 and the other one not.

7 130

8 **a** not bipartite has it has a cycle of odd length.

b not bipartite has it has 12 edges and the maximum number of edges of a bipartite graph of order 6 is 9.

c bipartite: $M = \{A, C, E\}$ and $N = \{B, D, F\}$ is a partition of the vertices of this graph.

d not bipartite has it has a cycle of odd length.

10 **a**

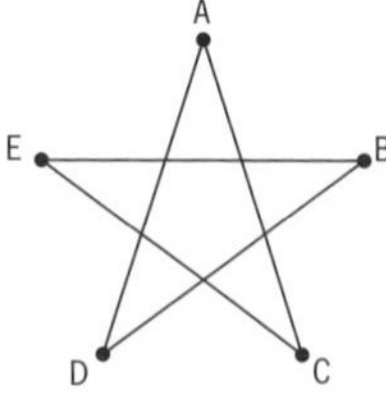

complement of C_5

The complement of C_5 is isomorphic to C_5

$A \to A$, $B \to C$, $C \to E$, $D \to B$, and $E \to D$, defines and isomorphism between C_5 and its complement.

c C_3 is isomorphic to K_3; W_3 is isomorphic to K_4

12 No wheel graph W_n, $n \geq 3$ can be bipartite because one its vertices is connected to all other vertices and each of the other vertices is connected to other vertices of the outside cycle of the wheel graph.

13 **b** W_n, $n \geq 3$ cannot contain an Eulerian trail because at least 3 of its vertices have odd degree 3. W_n, $n \geq 3$ cannot contain an Eulerian circuit either as it contain vertices of odd degree.

14 **a** The graph is planar as it has the following planar embedding:

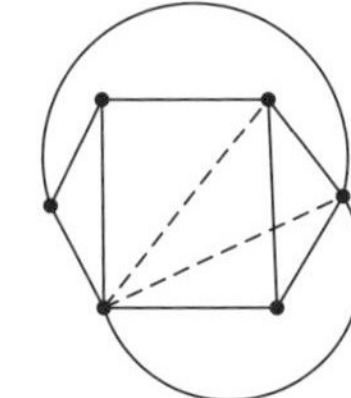

b The graph is planar as it consists of two isomorphic connected components with the following planar embedding:

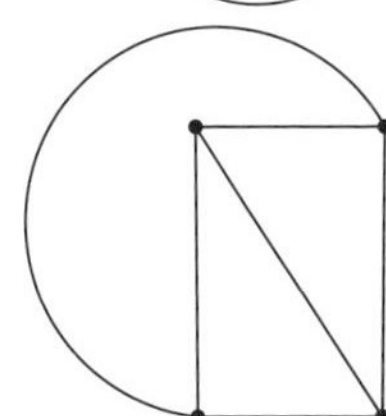

c This graph is not planar as it has the following subgraph isomorphic to K_5:

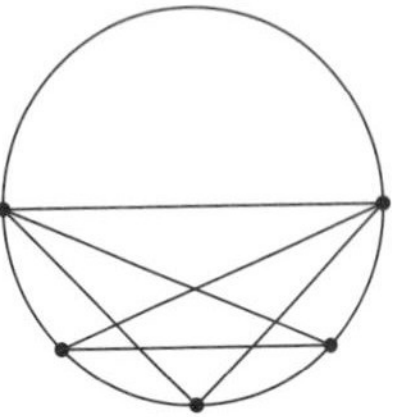

d This graph is not planar as it has the following subgraph isomorphic to $K_{3,3}$:

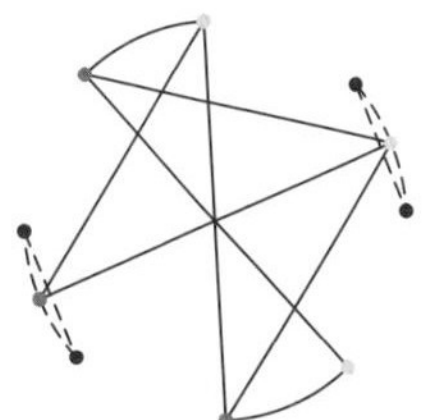

e The graph is planar as it has the following planar embedding:

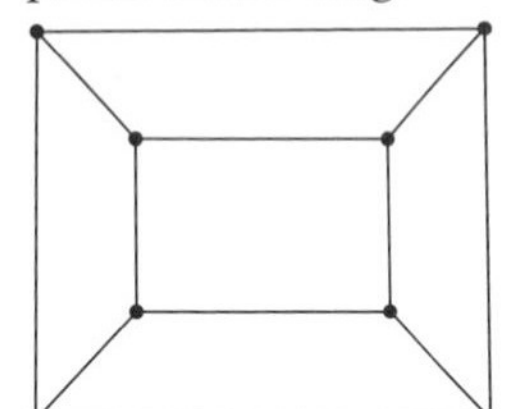

f The graph is planar as it has the following planar embedding:

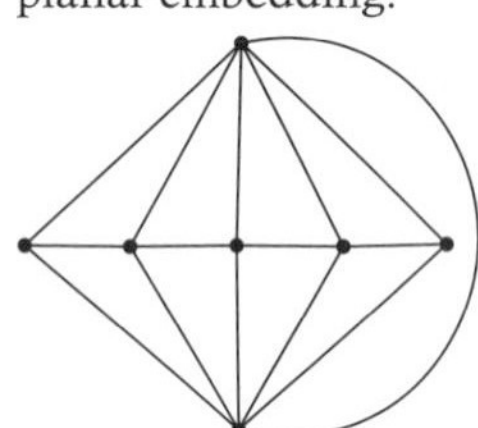

15 22

16 7

17 **a**

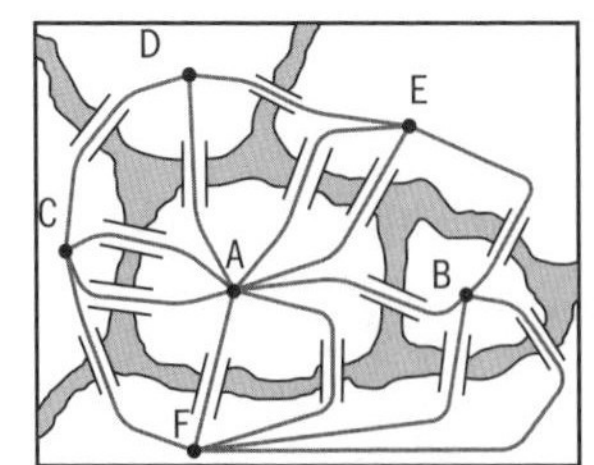

b

	A	B	C	D	E	F
A	0	1	2	1	2	2
B	1	0	0	0	1	2
C	2	0	0	1	0	1
D	1	0	1	0	1	0
E	2	1	0	1	0	0
F	2	2	1	0	0	0

c deg (A) = 8; deg (B) = 4; deg (C) = 4; deg (D) = 3; deg (E) = 4 and deg (F) = 5

d i This graph has no Eulerian circuit because it contains vertices of odd degree;

ii As it contain exactly two vertices of odd degree (D and F), it is possible to find a Eulerian trail starting and ending at these vertices.

20 b 5

Chapter 5

Skills check

1 a

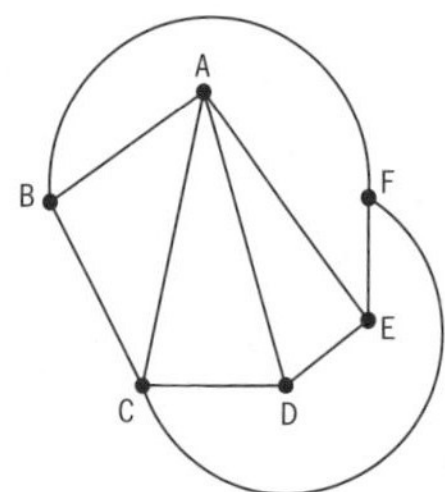

not Eulerian; Hamiltonian.

b

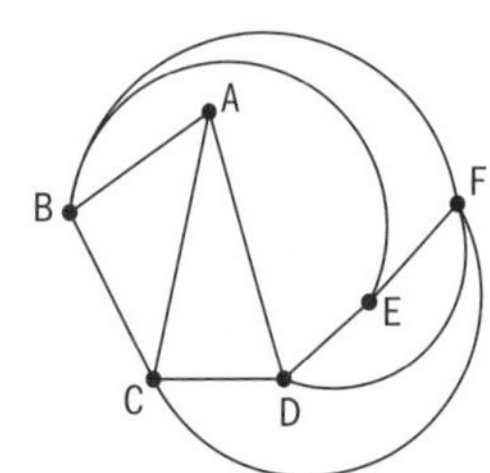

not Eulerian; Hamiltonian.

2 This graph 6 vertices; a spanning tree has $6 - 1 = 5$ edges. Here are 5 possible spanning trees of this graph:

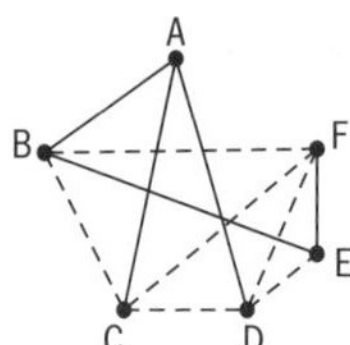

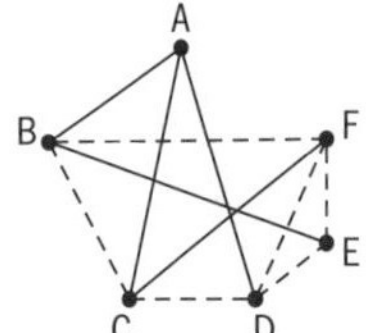

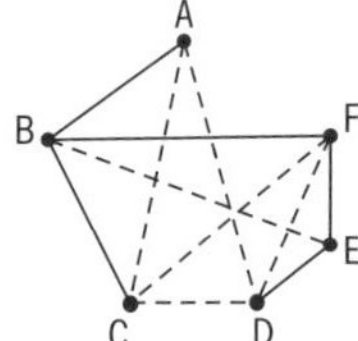

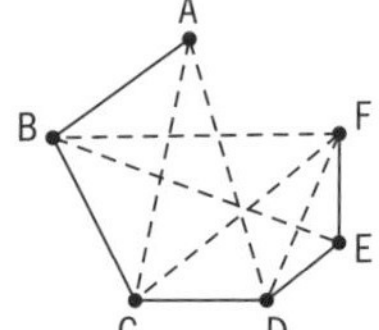

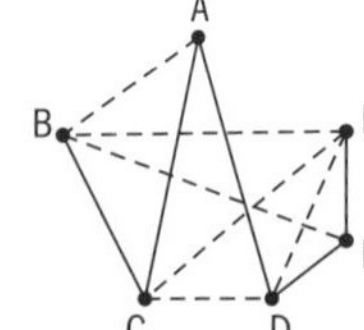

Exercise 5A

1 a

	A	B	C	D	E
A	–	6	8	10	7
B	6	–	11	6	12
C	8	11	–	14	–
D	10	6	14	–	5
E	7	12	–	5	–

b

	A	B	C	D	E
A	–	6	3	2	7
B	6	–	4	–	–
C	3	4	–	8	9
D	2	–	8	–	5
E	7	–	9	5	–

c

	A	B	C	D	E	F
A	–	6	–	2	7	9
B	6	–	4	–	–	4
C	–	4	–	8	–	–
D	2	–	8	–	5	9
E	7	–	–	5	–	8
F	9	4	–	9	8	–

d

	A	B	C	D	E	F	G
A	–	6	–	2	–	–	3
B	6	–	4	–	–	4	–
C	–	4	–	8	4	–	–
D	2	–	8	–	5	9	7
E	–	–	4	5	–	5	–
F	–	4	-	9	5	–	6
G	3	–	–	7	–	6	–

2 a Start the tree T with the edge DE with weight 5, then select AB and BD with weights 6 and finally AC with weight 8.
The total weight of the MST is 25.

b Start the tree T with the edge AD with weight 2, then select AC with weight 3, BC with weight 4 and finally DE with weight 5.
The total weight of the MST is 14.

c Start the tree T with the edge AD with weight 2, then select BC and BF with weights 4, then DE with weight 5 and finally AB with weight 6.
The total weight of the MST is 21.

d Start the tree T with the edge AD with weight 2, then AG with weight 3, then select BC, BF and CE with weight 4 and finally DE with weight 5.
The total weight of the MST is 22.

3 a i

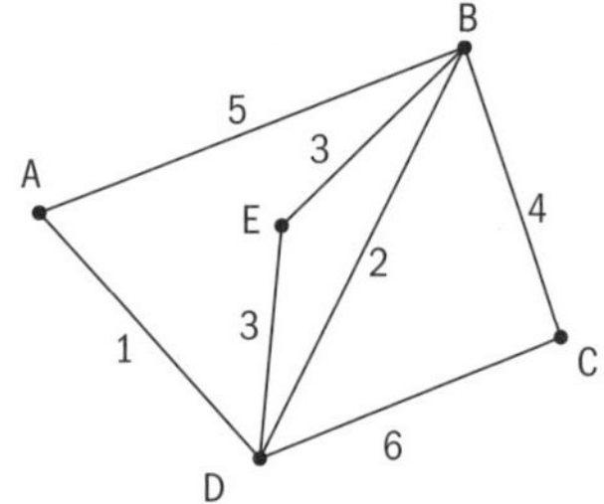

ii Start the tree T with the edge AD with weight 1, then select BD with weight 2, then BE with weight 3 and finally BC with weight 4.

iii The total weight of the MST is 10.

b i

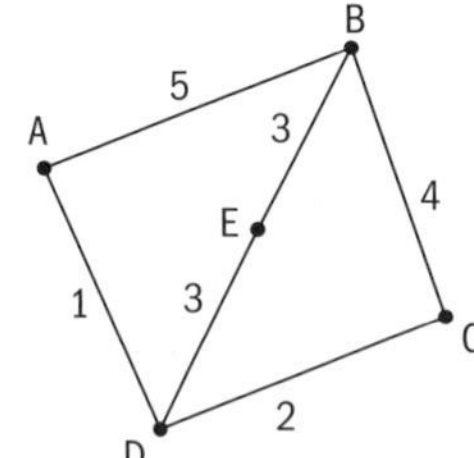

ii Start the tree T with the edge AD with weight 1, then select DC with weight 2 and finally BE and DE with weights 3.

iii The total weight of the MST is 9.

c i

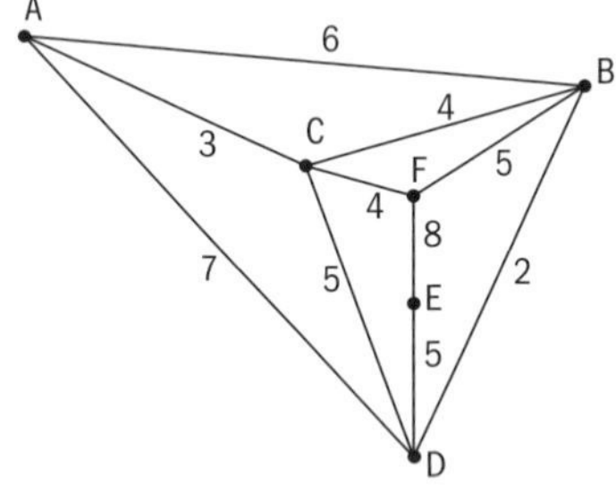

ii Start the tree T with the edge BD with weight 2, then select AC with weight 3, then BC and CF, each with weight 4, and finally DE with weight 5.

iii The total weight of the MST is 18.

d i

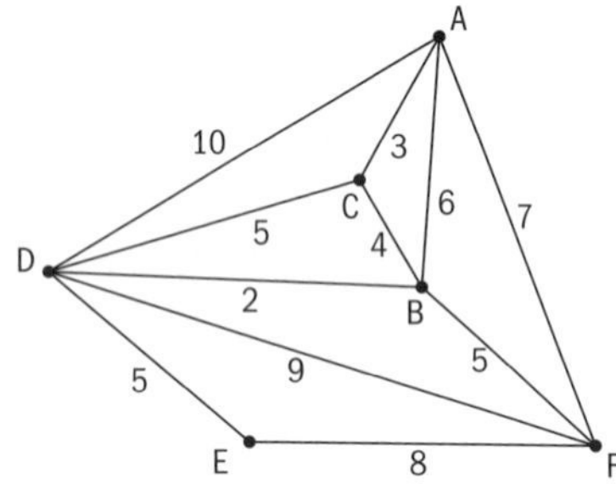

ii Start the tree T with the edge BD with weight 2, then select AC with weight 3, then BC with weight 4 and finally BF and DE with weights 5.

iii The total weight of the MST is 19.

5 56 **6** 6 **7** 17.88 euros

8 Kruskal's algorithm does not provide a solution to visit all the rooms as the solution is not an Hamiltonian path. Katharina can score 50 points if she starts at G and visits G, A, C, I and F or if she starts at H and visits H, E, D, B and A. Note that you can reverse the order of these paths to obtain other options.

Exercise 5B

1 a

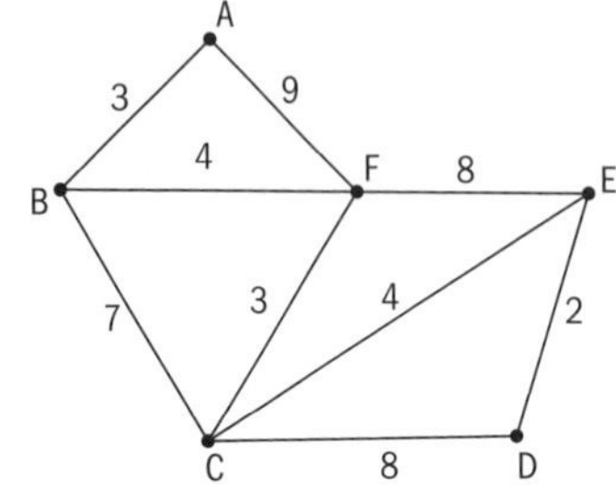

b

Step	A	B	C	D	E	F	vertex added
1	0	3 (A)	∞	∞	∞	9 (A)	**A**
2	…	…	10 (B)	∞	∞	7 (B)	**B**
3	…	…	10 (B)	∞	15 (F)	…	**F**
4	…	…	…	18 (C)	14 (C)	…	**C**
5	…	…	…	16 (E)	…	…	**E**
	…	…	…	…	…	…	**D**

The shortest path from A to D has length 16.

2 a Kruskal's algorithm (to find the MST)

b To obtain the MST for this graph add the edges CD, CF, EF, AC and BC. This tree has total weight 29 (i.e. the minimum cost of this road system is 29 hundred million euros).

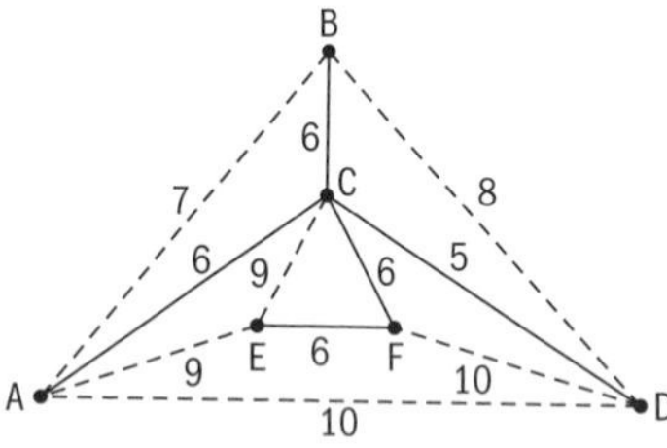

3 a

	A	B	C	D	E	F
A	–	9	–	–	–	8
B	9	–	2	3	4	–
C	–	2	–	4	–	–
D	–	–	4	–	6	7
E	–	4	–	6	–	8
F	8	–	–	7	8	–

b Add the edges BC, BD, BE, DF and AF. The MST has weight 24.

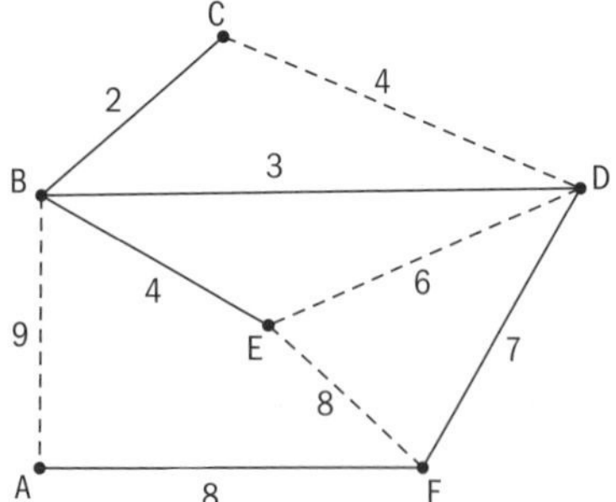

c i The graph contain triangles. Therefore it cannot be bipartite.

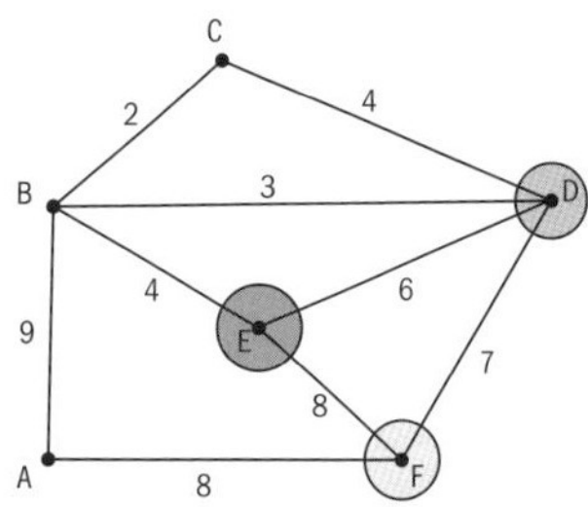

ii The graph is semi-Eulerian because it has exactly two vertices with odd degree: E and F. An Eulerial trail is EBCDBAFEDF.

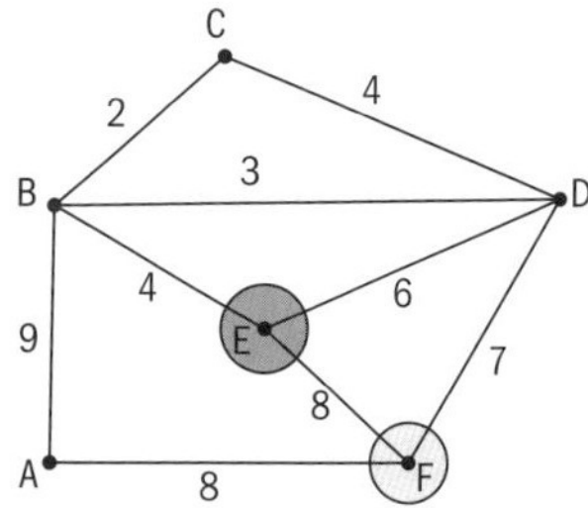

Exercise 5C

1 a Graph R: B and E; graph S: B and E; graph T: B, C ,D and F

b Graphs 1 and 2 have two vertices of odd degree and graph 3 has 4 vertices of odd degrees. Therefore these graphs are not Eulerian.

c For graphs 1 and 2 there is just a pair of vertices with odd degree. Therefore it is enough to duplicate the edges along the shortest path between these vertices as shown in the diagrams below. For graph 3 we need to inspect all possible combinations of pairing of vertices of odd order:

B and F, and C and D

B and C, and D and F

B and D, and C and F

For each case find by inspection the length of the shortest path between the pair of vertices:

B and F: 2, and C and D: 7. Therefore this pairing has total weight 9

B and C: 5, and D and F: 8. Therefore this pairing has total weight 13

B and D: 10, and C and F: 4. Therefore this pairing has total weight 14

Therefore the best pairing option has total weight 9 as shown in the diagram below.

	Graph 1	Graph 2	Graph 3
Diagram showing additional edges			
Total weight of all the edges:	25	16	39
Weight of additional edges	7	2 + 3 = 5	2 + 7 = 9
Total weight of the shortest path:	25 + 7 = 32	16 + 2 + 3 = 21	39 + 2 + 7 = 48

2

3 30

Exercise 5D

1 a 26 **b** 30

2 a 68 **b** 58

3 a order 5 and size 10; complete graph

b 12

c i

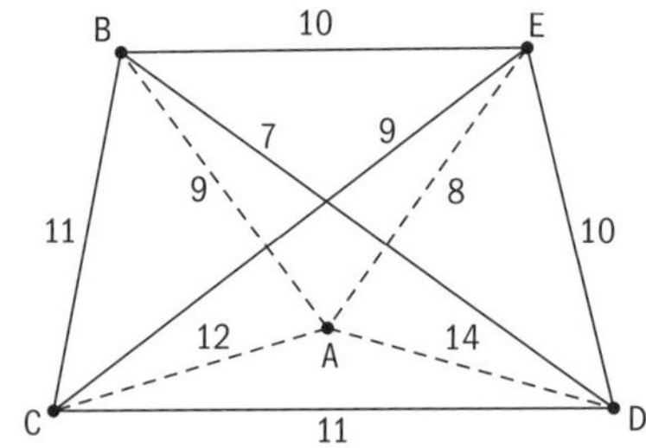

This tree has weight 26.

ii 43

d 46

Review Exercise

1 a

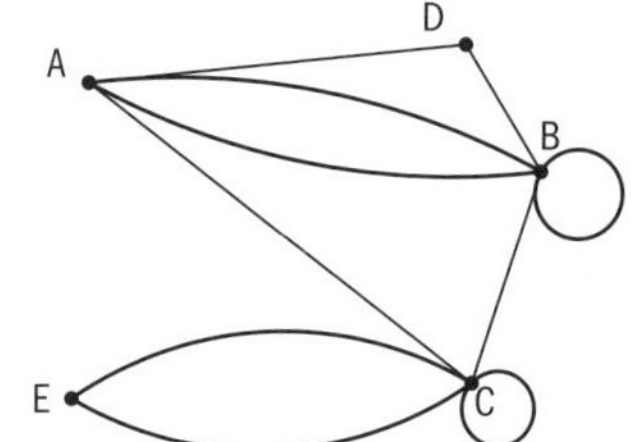

b i an Eulerian circuit is one that contains every edge of the graph exactly once

ii a possible Eulerian circuit is $A \to D \to B \to B \to C \to C \to E \to C \to A \to B \to A$

c i A Hamiltonian cycle passes through each vertex of the graph exactly once

ii To pass through E, you must have come from C and must return to C. Hence there is no Hamiltonian cycle for *G*.

2 a i The vertices B and D have odd degrees. Therefore there is no Eulerian circuit for this graph but it is possible to find a trail with endpoints at the vertices B and D.

ii $B \rightarrow A \rightarrow F \rightarrow B \rightarrow C \rightarrow F \rightarrow E \rightarrow C \rightarrow D \rightarrow A \rightarrow E \rightarrow D$

b i Add AD, AB, DC, CF and CE.

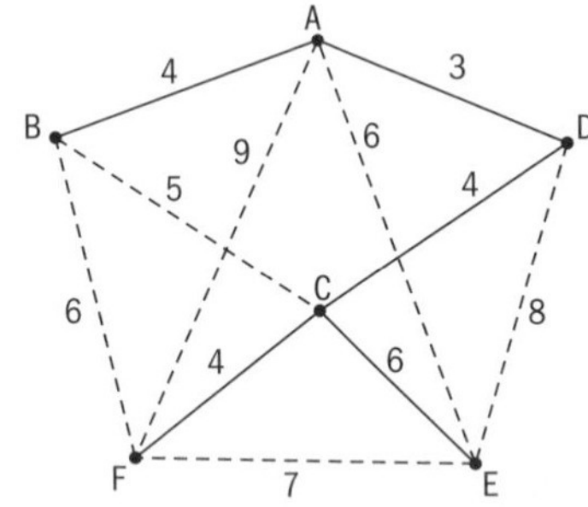

ii The weight of this spanning tree is 21.

c

Step	A	B	C	D	E	F	vertex added
1	0	4 (A)	∞	3 (A)	6 (A)	9 (A)	A
2	...	4 (A)	7 (D)	...	6 (A)	9 (A)	D
3	...	...	7 (D)	...	6 (A)	9 (A)	B
4	...	...	7 (D)	...	...	9 (A)	E
5	...	...	...	...	...	9 (A)	C
	...	...	...	...	...	...	F

3 a

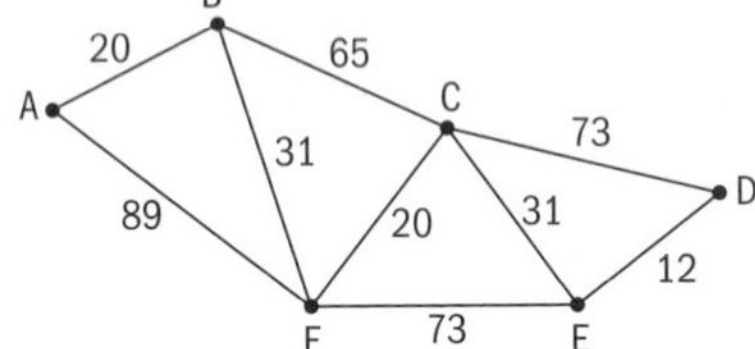

b

Step	A	B	C	D	E	F	vertex added
1	0	20 (A)	∞	∞	∞	89 (A)	A
2	...	...	85 (D)	∞	∞	51 (B)	B
3	...	...	71 (F)	∞	124 (F)	...	F
4	...	...	...	158 (C)	116 (F)	...	C
5	...	...	...	128 (E)	...	...	E
	...	...	...	...	...	...	D

Shortest path has length 128.

4 a

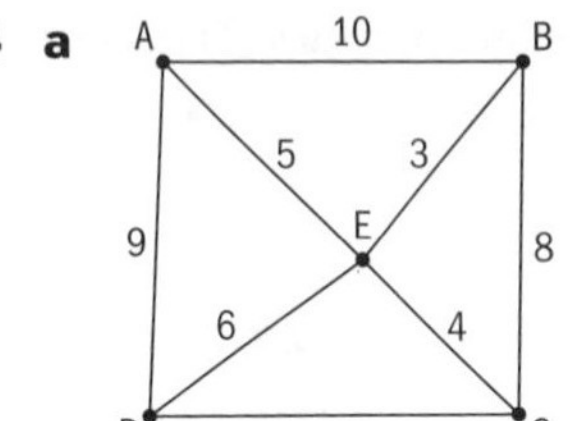

b The distinct Hamiltonian cycles are

ABCDEA

ABCEDA

ABECDA

AEBCDA

The weights are 37, 37, 34, 33 respectively.

The Hamiltonian cycle of least weight is AEBCDA.

c For a simple planar graph containing triangles, $e \leq 3v - 6$. Here $v = 5$, so $e \leq 9$. There are already 8 edges so the maximum number of edges that could be added is 1. This can be done *eg* AC or BD

5 a One upper bound is the length of any cycle, *eg* PTQRSP gives 120.

b i Using Kruskal's algorithm, the edges are introduced in the order PQ, PS and QR.

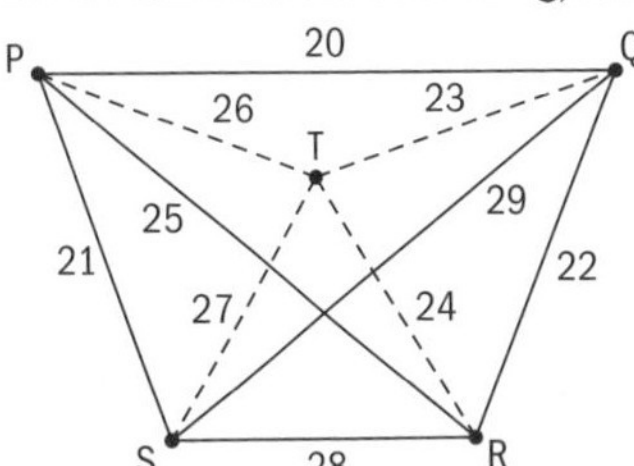

ii 108

6 a As there are 10 vertices we have 9 choices.

Choice	Edge	Weight
1	IG	1
2	JE	2
2	JF	2
4	ED	3
4	AI	3
6	DK	4
7	AB	5
7	FG	5
9	BC	6

Total weight = 31

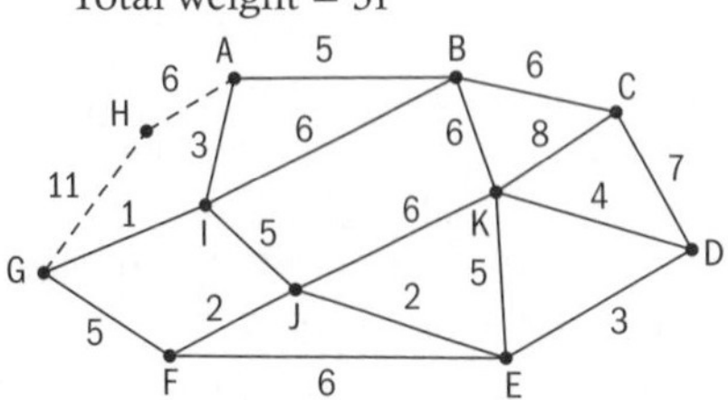

b 48

d $\frac{(11-1)!}{2} = 1814400.$

Index

A

adjacency tables 105
  cost adjacency tables 143–4
algorithms 141
  Chinese Postman algorithm 151–2
  deleted vertex algorithm for lower bound 155–9
  Dijkstra's Algorithm 146–9
  Euclidean algorithms 16–19
  greedy algorithms 142–6
  Kruskal's Algorithm 142–6
  Nearest Neighbour algorithm for upper bound 154
  search algorithms 146
amortizations 90–1

B

bases 4–13
binary code 10–11
Binet, Jacques 80
Binet's formula 80
Brahmagupta 4, 61
Buckyballs 124

C

chemical compound models 108
Chinese Postman Problem 149–50
  Chinese Postman algorithm 151–2
Chinese Remainder Theorem 57–63
circuits 109
  Eulerian circuits and trails 129–33
Clay Mathematics Institute 27
co-prime integers 28
composite integers 26
compound interest 91
computer science 111
congruences 44
  congruence modulo *n* 42–8
  modular inverses and linear congruences 48–53
  systems of linear congruences 57–63
constant coefficients 94–8
converting miles into kilometers 101
cost adjacency tables 143–4
counting problems 83–9
cryptography 41
cycles 109
  Eulerian circuits and trails 129–33
  Hamiltonian cycles 126–9
  length of a cycle 109
  using cycles for powers modulo *n* and Fermat's Little Theorem 64–71

D

deleted vertex algorithm for lower bound 155–9
denary (binary) 10–11
digraphs 109
Dijkstra, Edsger 146
Dijkstra's Algorithm 146–9
  limitation of 148
Diophantus of Alexandria 20
  Diophantus Riddle 20
  linear Diophantine equations 21–6
Dirac, Gabriel Andrew 127–8
Dirichlet, Gustav Lejeune 53
division theorem 15–16
  greatest common divisor 16
divisors 13–30

E

Eratosthenes 26
Euclid's Lemma 33
Euclid's statement 28–30
Euclidean algorithms 16–19
Euler relation for planar graphs 120–3
Euler, Leonhard 103, 121
Eulerian circuits and trails 129–33
  Chinese Postman Problem 149–52

F

factors 13–30
Fermat, Pierre de 68
Fermat's Last Theorem 53
Fermat's Little Theorem 69–71
Fibonacci 78
  Fibonacci numbers 79–80
  Fibonacci sequence 94
financial problems 89–90
first-degree linear recurrence relations 83–9
  general solutions of first-degree recurrence relations 100
  homogeneous relations 100
  modelling with first-degree recurrence relations 89–93
forests 111
four-colour theorem 141
Fundamental Theorem of Arithmetic 33–7
  alternative direct proof 35–6
  every positive integer *n* greater than 1 can be written uniquely as a product of primes 34

G

games 92–3
Gauss, Carl Friedrich 41
Germain, Sophie 42
Golden Ratio 101
Golden sequence 79–80
graph theory 102–3, 140
  algorithms and methods 141
  Chinese postman problem 149–52
  classification of graphs 104–8, 108–15
  deleted vertex algorithm for lower bound 155–9
  different representations of the same graph 115–17
  Eulerian circuits and trails 129–33
  Hamiltonian cycles 126–9
  minimum connector problems 142–6
  Nearest Neighbour algorithm for upper bound 154
  planar graphs 118–26
  shortest path problems 146–9
  terminology 104–8
  Travelling Salesman Problem 153–4
graphs 104–8
  2-colourable graphs 114
  adjacent vertices 104
  bipartite graphs 113–15
  complements of graphs 119–20
  complete bipartite graphs 114
  complete graphs 112–13
  connected graphs 110–11
  cost adjacency tables 143–4
  criterion for planarity 122
  definition 104
  degree of vertex 105
  directed graphs 109
  disconnected graphs 110
  edgeless graphs 120
  elementary subdivisions 122
  Euler relation for planar graphs 120–3
  forests 111
  Hamiltonian graphs 128
  Handshaking lemma 106–7
  homeomorphic graphs 122
  incident vertices 104
  isomorphism invariants 116–17
  minimum spanning tree 142

graphs (*continued*)
multiedges and loops 105
null graphs 120
order and size 104
partitions 113
planar graphs 118–26
regular graphs 118
representations 115–17
self-complementary graphs 120
semi-Hamiltonian graphs 128
shortest path tree 146
simple graphs 109
spanning trees 119
subgraphs 107
trees 111–12
weighted graphs 108
greedy algorithms 142–6

H

Hamilton, Sir William Rowan 126
Hamiltonian cycles 126–9
Handshaking lemma 106
corollary 107
Hilbert, David 21, 27

I

integers 13–26
prime numbers 26–30
investments 91
isomorphism invariants 116–17

K

Königsberg bridges problem 103, 129–33
Kroneker, Leopold 13
Kruskal, Joseph Bernard 142
Kruskal's Algorithm 142–6
Kuan Mei-Ko 149
Kuratowski, Kasimierz 122–3

L

Leonardo de Pisa 78
linear congruences 48–53
systems of linear congruences 57–63
linear Diophantine equations 21–6
corollary 22
loans 90–1
Lucas numbers 98
Lucas, François Édouard Anatole 98

M

Mersenne, Marin 29
modular arithmetic 40
Chinese Remainder Theorem 57–63
congruence modulo *n* 42–8
from Gauss to cryptography 41–2
modular inverses and linear congruences 48–53
Pigeonhole Principle 53–6
using cycles for powers modulo *n* and Fermat's Little Theorem 64–71
modular inverses 48–53
modulo *n* 42–8, 64–9
multigraphs 105
multiples 13
least common multiples 35–7

N

Nearest Neighbour Algorithm for upper bound 154
number systems 2
Fundamental Theorem of Arithmetic and least common multiples 33–7
history of number systems 3–4
integers, prime numbers, factors and divisors 13–30
number systems and bases 4–13
Rules of Brahmagupta 4
strong mathematical induction 30–3

O

octal (Base 8) 11

P

paths 109
degree sequences 109
perfect numbers 29
Pigeonhole Principle 53–6
planar graphs 118–26
Euler relation for planar graphs 120–3
faces 118
planar embedding 118
Plato 118
prime integers 26
prime numbers 26–30
Mersenne primes 29
probability problems 92–3

R

recurrence relations 78–83
definition 80
modelling with first-degree recurrence relations 89–93
second-degree linear homogeneous recurrence relations with constant coefficients 94–8
solution of first-degree linear recurrence relations and applications to counting problems 83–9
recursive patterns 76
games and probability problems 92–3
investments and compound interest 91
loans and amortizations 90–1
modelling and solving problems using sequences 77
modelling with first-degree recurrence relations 89–93
recurrence relations 78–83
second-degree linear homogeneous recurrence relations with constant coefficients 94–8
solution of first-degree linear recurrence relations and applications to counting problems 83–9
reflexive 44, 45
relatively prime integers 28
remainders 43
Chinese Remainder Theorem 57–63
Riemann Hypothesis 27
Riemann, Bernhard 27
RSA encryption 41

S

second-degree linear homogeneous recurrence relations 94–8
auxiliary equation 94
general solutions of second-degree homogeneous recurrence relations 101
shortest path problems 146–9
sieve of Eratosthenes 26
strong mathematical induction 30–3
subgraphs 107
Sun Zi Suanjing *The Mathematical Classic of Sun Zi* 57
symmetric 44, 45

T

Tower of Hanoi 85–6
trails 109
Eulerian circuits and trails 129–33
transitive 45
Travelling Salesman Problem 153–4
brute force method 153
trees 111–12
minimum spanning tree 142
shortest path tree 146
spanning trees 119

V

value of N in base b 6

W

walks 109
weak mathematical induction 30
well-ordered relations 14–15
Wiles, Andrew 68

X

Xunyu Zhou 77